SPACE SCIENCE IN THE TWENTY-FIRST CENTURY: IMPERATIVES FOR THE DECADES 1995 TO 2015

FUNDAMENTAL PHYSICS AND CHEMISTRY

Task Group on Fundamental Physics and Chemistry
Space Science Board
Commission on Physical Sciences, Mathematics, and Resources
National Research Council

NATIONAL ACADEMY PRESS
Washington, D.C. 1988

National Academy Press • 2101 Constitution Avenue, N.W. • Washington, D. C. 20418

NOTICE: The project that is the subject of this report was approved by the Governing Board of the National Research Council, whose members are drawn from the councils of the National Academy of Sciences, the National Academy of Engineering, and the Institute of Medicine. The members of the committee responsible for the report were chosen for their special competences and with regard for appropriate balance.

This report has been reviewed by a group other than the authors according to procedures approved by a Report Review Committee consisting of members of the National Academy of Sciences, the National Academy of Engineering and the Institute of Medicine.

The National Academy of Sciences is a private, nonprofit, self-perpetuating society of distinguished scholars engaged in scientific and engineering research, dedicated to the furtherance of science and technology and to their use for the general welfare. Upon the authority of the charter granted to it by the Congress in 1863, the Academy has a mandate that requires it to advise the federal government on scientific and technical matters. Dr. Frank Press is president of the National Academy of Sciences.

The National Academy of Engineering was established in 1964, under the charter of the National Academy of Sciences, as a parallel organization of outstanding engineers. It is autonomous in its administration and in the selection of its members, sharing with the National Academy of Sciences the responsibility for advising the federal government. The National Academy of Engineering also sponsors engineering programs aimed at meeting national needs, encourages education and research, and recognizes the superior achievements of engineers. Dr. Robert M. White is president of the National Academy of Engineering.

The Institute of Medicine was established in 1970 by the National Academy of Sciences to secure the services of eminent members of appropriate professions in the examination of policy matters pertaining to the health of the public. The Institute acts under the responsibility given to the National Academy of Sciences by its congressional charter to be an adviser to the federal government and, upon its own initiative, to identify issues of medical care, research, and education. Dr. Samuel O. Thier is president of the Institute of Medicine.

The National Research Council was organized by the National Academy of Sciences in 1916 to associate the broad community of science and technology with the Academy's purposes of furthering knowledge and advising the federal government. Functioning in accordance with general policies determined by the Academy, the Council has become the principal operating agency of both the National Academy of Sciences and the National Academy of Engineering in providing services to the government, the public, and the scientific and engineering communities. The Council is administered jointly by both Academies and the Institute of Medicine. Dr. Frank Press and Dr. Robert M. White are chairman and vice chairman, respectively, of the National Research Council.

Support for this project was provided by Contract NASW 3482 between the National Academy of Sciences and the National Aeronautics and Space Administration.

Library of Congress Catalog Card Number 87-43331

ISBN 0-309-03841-3

Printed in the United States of America

TASK GROUP ON FUNDAMENTAL PHYSICS AND CHEMISTRY

Rainer Weiss, Massachusetts Institute of Technology,
 Co-Chairman
Joseph M. Reynolds, The Louisiana State University,
 Co-Chairman
Peter Bender, University of Colorado
A. L. Berlad, University of California
Russell Donnelly, University of Oregon
Freeman Dyson, The Institute of Advanced Study
William M. Fairbank, Stanford University
Robert Hofstadter, Stanford University
George Homsy, Stanford University
James Langer, University of California
John E. Naugle, Consultant, Chevy Chase, Maryland
Rene Pellat, CNES
Remo Ruffini, Universita di Roma
Dudley Saville, Princeton University
John Robert Schrieffer, University of California

Dean P. Kastel, *Staff Director*
Ceres M. Rangos, *Secretary*

STEERING GROUP

Thomas M. Donahue, University of Michigan, Chairman
Don L. Anderson, California Institute of Technology
D. James Baker, Joint Oceanographic Institutions, Inc.
Robert W. Berliner, Pew Scholars Program, Yale University
Bernard F. Burke, Massachusetts Institute of Technology
A. G. W. Cameron, Harvard College Observatory
George B. Field, Center for Astrophysics, Harvard University
Herbert Friedman, Naval Research Laboratory
Donald M. Hunten, University of Arizona
Francis S. Johnson, University of Texas at Dallas
Robert Kretsinger, University of Virginia
Stamatios M. Krimigis, Applied Physics Laboratory
Eugene H. Levy, University of Arizona
Frank B. McDonald, NASA Headquarters
John E. Naugle, Chevy Chase, Maryland
Joseph M. Reynolds, The Louisiana State University
Frederick L. Scarf, TRW Systems Park
Scott N. Swisher, Michigan State University
David A. Usher, Cornell University
James A. Van Allen, University of Iowa
Rainer Weiss, Massachusetts Institute of Technology

Dean P. Kastel, *Study Director*
Ceres M. Rangos, *Secretary*

SPACE SCIENCE BOARD

Thomas M. Donahue, University of Michigan, Chairman
Philip H. Abelson, American Association for the Advancement of Science
Roger D. Blandford, California Institute of Technology
Larry W. Esposito, University of Colorado
Jonathan E. Grindlay, Center for Astrophysics
Donald N. B. Hall, University of Hawaii
Andrew P. Ingersoll, California Institute of Technology
William M. Kaula, NOAA
Harold P. Klein, The University of Santa Clara
John W. Leibacher, National Solar Observatory
Michael Mendillo, Boston University
Robert O. Pepin, University of Minnesota
Roger J. Phillips, Southern Methodist University
David M. Raup, University of Chicago
Christopher T. Russell, University of California, Los Angeles
Blair D. Savage, University of Wisconsin
John A. Simpson, Enrico Fermi Institute, University of Chicago
George L. Siscoe, University of California, Los Angeles
L. Dennis Smith, Purdue University
Darrell F. Strobel, Johns Hopkins University
Byron D. Tapley, University of Texas at Austin

Dean P. Kastel, *Staff Director*
Ceres M. Rangos, *Secretary*

COMMISSION ON PHYSICAL SCIENCES, MATHEMATICS, AND RESOURCES

Norman Hackerman, Robert A. Welch Foundation, *Chairman*
George F. Carrier, Harvard University
Dean E. Eastman, IBM Corporation
Marye Anne Fox, University of Texas
Gerhart Friedlander, Brookhaven National Laboratory
Lawrence W. Funkhouser, Chevron Corporation (retired)
Phillip A. Griffiths, Duke University
J. Ross Macdonald, University of North Carolina, Chapel Hill
Charles J. Mankin, Oklahoma Geological Survey
Perry L. McCarty, Stanford University
Jack E. Oliver, Cornell University
Jeremiah P. Ostriker, Princeton University Observatory
William D. Phillips, Mallinckrodt, Inc.
Denis J. Prager, MacArthur Foundation
David M. Raup, University of Chicago
Richard J. Reed, University of Washington
Robert E. Sievers, University of Colorado
Larry L. Smarr, National Center for Supercomputing Applications
Edward C. Stone, Jr., California Institute of Technology
Karl K. Turekian, Yale University
George W. Wetherill, Carnegie Institution of Washington
Irving Wladawsky-Berger, IBM Corporation

Raphael G. Kasper, *Executive Director*
Lawrence E. McCray, *Associate Executive Director*

Foreword

Early in 1984, NASA asked the Space Science Board to undertake a study to determine the principal scientific issues that the disciplines of space science would face during the period from about 1995 to 2015. This request was made partly because NASA expected the Space Station to become available at the beginning of this period, and partly because the missions needed to implement research strategies previously developed by the various committees of the board should have been launched or their development under way by that time. A two-year study was called for. To carry out the study the board put together task groups in earth sciences, planetary and lunar exploration, solar system space physics, astronomy and astrophysics, fundamental physics and chemistry (relativistic gravitation and microgravity sciences), and life sciences. Responsibility for the study was vested in a steering group whose members consisted of the task group chairmen plus other senior representatives of the space science disciplines. To the board's good fortune, distinguished scientists from many countries other than the United States participated in this study.

The findings of the study are published in seven volumes: six task group reports, of which this volume is one, and an overview report of the steering group. I commend this and all the other task group reports to the reader for an understanding of the challenges that confront the space sciences and the insights they promise for the next century. The official recommendations of the study are those to be found in the steering group's overview.

Thomas M. Donahue, Chairman
Space Science Board

Preface

The efforts of the Task Group on Fundamental Physics and Chemistry have been directed toward an understanding of the role of the space program in furthering our knowledge of the fundamental interactions in nature. All of scientific inquiry, of course, has some bearing on this. However, this task group's emphasis has been on direct experiments designed to enhance this fundamental knowledge. The focus of the group has been on research in two principal areas: (1) relativistic gravitation, which involves tests of the general theory of relativity, and (2) microgravity science, which encompasses experiments in the low-acceleration and low-gravity gradient environments of space. Early on, a subgroup for each area was formed to identify opportunities in these two fields. The findings of these subgroups make up the two parts of this report. It should be noted that the task group decided at the outset not to study applied microgravity science and the utility of space for manufacturing and industrial processes.

This study of the status of and prospects for gravitational physics in the space program has benefited from prior studies made by committees of both NASA and the National Research Council. Many elements of the proposed program have been recommended in the following: Report of the Sub-panel on Relativity and Gravitation MOWG in Shuttle Astronomy, NASA, 1976; *Strategy for Space Research in Gravitational Physics in the 1980s*, National Academy Press, Washington, D.C., 1981; and *Survey of Gravitation, Cosmology, and Cosmic Ray Physics*, Physics Survey Committee, National Academy Press, Washington, D.C., 1986.

Contents

A. GRAVITATIONAL PHYSICS

1. INTRODUCTION   3
2. BASIC ISSUES IN GRAVITATION   8
   Tests of Fundamental Principles, 8
   Weak Field, 9
   Strong Field, 11
   Nonstatic Fields, 13
   Cosmology, 15
3. GRAVITATIONAL WAVE ASTRONOMY   21
4. CURRENT SPACE RESEARCH IN GRAVITATION   26
   Lunar Ranging, 26
   Analysis of Planetary and Lunar Motion, 27
5. EXPECTED RESEARCH PRIOR TO 1995   30
   Shuttle Test and Flight of Gravity Probe B (GPB), 30
   Shuttle Flight of a Cryogenic Principle of Equivalence Experiment, 31
   Microwave Ranging to the Mars Observer Spacecraft, 33
   X-ray Timing Experiment and a Medium-Area Fast X-ray Detector on the Shuttle, 34

Spacecraft Observations of Long-Period Gravitational Waves, 36

6. PROGRAMS AFTER 1995   38
 Baseline Program, 38
  LAGOS: Laser Gravitational Wave Observations in Space, 38
  POINTS: Small Astrometric Interferometer in Space, 44
  Mercury Relativity Satellite, 45
  STARPROBE: Second-Order Gravitational Red Shift Experiment, 49
  High-Precision Principle of Equivalence Experiment, 50
  X-ray Large Array/Fast-timing Experiments and Correlation with Gravitational Radiation, 51
 Enhanced Program, 51
  Laser Gravitational Wave Observatory, 51
  Reflight of GPB, 53

7. TECHNOLOGY REQUIREMENTS   55
 Improved Disturbance Compensation Systems (DISCOS), 55
 Moderate-Power Frequency-Stabilized Lasers, 55
 Cryogenic Capability to Transfer Helium in Orbit, 56
 Development of Clocks, 56

B. SCIENCE IN A MICROGRAVITY ENVIRONMENT

1. INTRODUCTION   59

2. GRAVITY-SENSITIVE SYSTEMS AT EQUILIBRIUM   66
 Critical Phenomena, 66
 Mechanics of Granular Media, 69

3. GRAVITATIONAL DESTABILIZATION OF STATIONARY STATES   71
 Mechanics of Suspensions, 71
 Sheared Suspensions of Granular Materials, 72
 Growth of Condensates in Supersaturated Systems, 74
 Fractal Aggregates, 75

4. SYSTEMS FAR FROM EQUILIBRIUM   77
 Solidification Patterns, 77

Surface Tension and Convection Effects, 79
  Minimizing Buoyancy-driven Effects, 79
  Minimizing Free Surface Deflections, 80
Minimizing the Effect of Gravity on Free Surface Shapes, 80
Electrokinetics, 81
Combustible Media, 83

5. SCALING AND ACCEPTABLE ACCELERATION LEVEL 86

6. DISCUSSION AND RECOMMENDATIONS 90

Part A
Gravitational Physics

1
Introduction

Gravitational physics grew out of astronomy in the seventeenth century. Newtonian gravitation, besides being an adequate and successful description of the gravitational interaction for everyday purposes, had a profound influence on the philosophy of natural science. It was the first example of the universality of physical laws; the force that caused Newton's apocryphal apple to fall was the same reason that the Moon falls toward the Earth, or the Earth toward the Sun. As has been known since the discovery of special relativity at the beginning of this century, Newtonian gravitation cannot be a complete theory of gravitation. It cannot be expressed in a relativistically covariant way and is not a field theory in the modern sense. The epoch between the formulation of special relativity in 1905 and the formulation of general relativity in 1916 witnessed the development of many theories of gravitation consistent with special relativity: scalar, vector, and tensor theories in four dimensions. A few of these theories survive, and several new theories have been developed in recent times. None carry the weight of general relativity, but they do serve to highlight the sensitive points in general relativity and help to generate a scheme for testing general relativity.

General relativity imbeds gravitation in a theory that relates the structure of space and time to the distribution of matter. As such, the framework set by general relativity becomes the arena for the other interactions in nature. Indeed, Einstein and others devoted considerable effort after the discovery of general relativity to an unsuccessful attempt at trying to unify, through geometric considerations alone, all the known interactions in nature. It is ironic that recent attempts at this grand unification from quantum field theory appear to be possible for all fields but gravitation. The failure is in good measure due to the inability, at present, to formulate a viable quantum theory of gravity—an outstanding theoretical problem.

Most of the consequences of classical (nonquantized) general relativity have been theoretically studied over the past 70 years. In its pristine form, the theory has no free parameters. A modification to the theory, noted by Einstein shortly after its first formulation, is the addition of a new term in the field equations, which would only have an effect on space-time on the largest scales, in particular in cosmological solutions.

In a purely mathematical sense the theory is complete and can be tested at any level. One could hold the extreme view that the agreement of general relativity with Newtonian gravitation in the low-velocity, weak-field limit is adequate empirical justification for the entire theoretical structure. A more critical viewpoint is that the failure of the theory in any test that can be devised is sufficient reason to render the entire theory invalid.

In the past 20 years, substantial theoretical progress has been made toward understanding the inner structure of relativistic gravitation. Extensive work has been done on formal solutions to the Einstein field equations and the meaning of the singularities in the theory. The properties of black hole solutions have been investigated in sufficient depth to make firm predictions concerning the gravitational radiation emitted by them in encounters with other black holes or stars. In the past 10 years, numerical relativity has become an active field by which the solutions of the nonlinear differential equations of general relativity can be visualized. A particularly important aspect of the numerical modeling is the coupling of general relativity with fluid dynamics and magnetohydrodynamics in the attempt to understand astrophysical phenomena such as stellar collapse, in which relativistic gravity may play a major role in the final stage. A technique has been developed for weak gravitational fields that expands metric theories of relativistic gravity, including general relativity, into a set of terms multiplied by parameters associated with the degree that a particular physical concept or symmetry is incorporated into the theory. This parametric post-Newtonian (PPN) expansion serves as a useful framework with which to test relativistic gravitation.

The profound consequences of general relativity are the phenomena that involve strong space-time curvature and the large-scale global properties of space-time. In these regimes, gravitation dominates all other interactions and Newtonian gravitation is not even a fair approximation of reality. The unambiguous discovery of a black hole, and especially the observation of the gravitational radiation due to the interaction of a black hole with its surroundings, would constitute the strongest confirmation of general relativity. A complete understanding of cosmic evolution would provide as sharp a test; however, the prospects for this are not as bright considering the complexity of interpreting the astrophysical phenomena that are the source of cosmological data.

General relativity makes specific predictions for new effects when the sources of the field are in motion. In analogy with electricity and magnetism, the gravitational field can be divided into electric and magnetic terms. In a weak field, the electric terms correspond to Newtonian gravitation. The new magnetic terms will come into evidence when both source and test body are in relative motion. The magnetic interaction between the spinning Earth and a spinning gyro is measurable and is the basis of a new test of general relativity to be attempted in the space program within the next 10 years. Magnetic gravitation is expected to be a large effect in the field of a spinning black hole. It may play a role in astrophysical phenomena, in particular in the explanation of the jets in extragalactic double radio sources. Gravitational radiation is predicted in all relativistic theories of gravitation. The sources of the radiation are accelerating masses. In general relativity the lowest-order radiation is due to the time-dependent quadrupole moment of the source. Binary stellar systems, aspheric stellar collapse, colliding stellar systems, and rotating neutron stars (pulsars) are examples of sources of gravitational radiation that may be detectable on Earth or by space-based receiving systems. The direct detection of the radiation could serve to test general relativity by confirming the propagation speed and polarization states of the waves. Furthermore, as already indicated, gravitational waves may be the best way to detect black holes and thereby test general relativity in the strong-field, high-velocity limit.

Gravitational wave astronomy could well open a new view of the universe as it will expose phenomena that involve the coherent motions of large masses at relativistic speeds. These processes will generally occur in the innermost parts of active astrophysical regions and therefore be obscured from "view" in the electromagnetic astronomies. A particularly exciting although highly speculative prospect is the possible detection of gravitation radiation noise from the very earliest epoch of the universe that is completely obscured in the electromagnetic and even neutrino astronomies.

The first observational evidence for relativistic gravitation was the small discrepancy in the orbital motion of the planet Mercury discovered in the nineteenth century. The orbit, taking account of the perturbations due to the other objects in the solar system, did not close on itself as predicted by Newtonian gravitation. Einstein was well aware of this and considered the ability of general relativity to predict the motion of Mercury as strong evidence for the correctness of the then new theory. The perihelion advance of Mercury as well as the bending of starlight by the Sun and the gravitational red shift of spectral lines emanating from the Sun relative to the same lines observed on Earth are the three classical tests suggested by Einstein. It is interesting to note that although the bending of light was observed in the well-publicized Eddington eclipse expedition of 1918, a reliable measurement of this effect had to wait for the radio interferometer observations of quasi-stellar sources passing near the Sun in the past decade. The gravitational red shift was not well measured until the middle 1960s by Mossbauer techniques on the ground and, more recently, by the comparison of two hydrogen maser clocks, one on the Earth's surface and the other in a suborbital trajectory. In the mid-1960s, a new test became possible with the emergence of radar astronomy that enabled the direct determination of the range to solar system bodies, thereby measuring directly the space-time curvature of the region around the Sun.

One of the most fortuitous discoveries for the observational tests of relativistic gravitation is the binary pulsar system, which was uncovered in a general radio search for pulsars. This system appears to be composed of two neutron stars, one of which is a stable pulsar and therefore a "good" clock. The kinematics of this system, observed through the Doppler shift of the pulsar signals as it orbits about its companion, has established several of the classical relativistic tests and furthermore is now demonstrating the first evidence for gravitational radiation damping.

Besides the red shift experiment mentioned above, the space program has played an important role in experimental tests of general relativity. Laser ranging to the corner reflectors placed on the Moon by the Apollo program has allowed a test of the principle of equivalence for gravitational self-energy. The Viking program through the Mars Orbiter and Lander has provided range information without the perturbations due to planetary topography. The precision timing of radio signals across the solar system to Viking and back to Earth has established the best observation of the post-Newtonian time delay in the vicinity of the Sun. The precision ephemerides of Mars, in concert with those of the other planets in the solar system, have set the best limits on many of the post-Newtonian parameters, including limits on the change of G (Newtonian gravitational constant) with time.

The common element in this historical perspective of progress in gaining empirical evidence for relativistic gravitation is the close coupling to technical advances in both direct experiments and observational techniques. There is no reason this should change in the future. The space program is expected to play a decisive role in this research.

2
Basic Issues in Gravitation

TESTS OF FUNDAMENTAL PRINCIPLES

The principle of equivalence led Einstein to the geometrization of gravity. The principle in its weakest form states that the ratio of the inertial mass—the tendency of a body to resist acceleration—to the passive gravitational mass—the tendency of a body to respond to a gravitational field—is the same for all bodies. This is a well-known result that is often trivialized in elementary mechanics courses. It is a profound statement when one considers that masses are not simple but composed of a host of different particles experiencing different interactions, each of which contributes to the mass. Gravity appears to be oblivious to this complexity. The ratio of the inertial to the gravitational mass is known from terrestrial experiments to be the same for all bodies to one part in 10^{11}. This implies that gravity does not differentiate between the strong and the electromagnetic interactions. Using laser ranging to the corner reflectors left on the Moon by the Apollo program, precision measurements of the motion of the Moon and Earth as they orbit the Sun have determined that the contribution to the mass of the gravitational binding energy of the Earth shares in this equivalence of the masses.

The equivalence of the masses leads directly to the inability to distinguish locally between a gravitational field and an acceleration: the principle of equivalence. It was through the application of special relativity to the kinematics and dynamics in a rotating system that Einstein was led to curved space and the geometric description of gravitation. The way that the principle of equivalence is incorporated in the general theory of relativity goes further than merely the equivalence of inertial and gravitational mass. The indistinguishability of gravity and acceleration implies that all physics in a local region experiencing free fall is the same whether free fall occurs near a massive object such as a black hole or far from other matter; in other words the gravitational potential does not enter local physics in any measurable way. The theory asserts the still bolder generalization that the formulation and numerical content of the laws of physics locally in a free falling region are independent of epoch and the physical conditions in the universe.

The principle of equivalence is at the foundation of gravitation. Violations of the principle would indicate the presence of new interactions in nature and would be evidence against the metric interpretation of gravity. New tests with increased accuracy are qualitatively different from prior measurements as they probe new regimes that could cause a violation of the principle. Space experiments in low g and g are expected to be able to improve the accuracy of the tests.

WEAK FIELD

A measure of the strength of the gravitational interaction is provided by the dimensionless constant GM/Rc^2, where G is the Newtonian gravitational constant, M the mass of the source body, R the distance to the observer, and c the velocity of light. On the surface of the Earth this quantity is 7×10^{-10}, while at the limb of the Sun it is 2×10^{-6}. The value at the location of the pulsar in the binary pulsar system is almost the same as at the limb of the Sun. These are all weak fields. On the surface of a neutron star it is close to 1×10^{-1}, and at the horizon of a black hole it nears unity. All of the detailed tests of relativistic gravitation have been carried out in fields where the constant is less than 10^{-6}.

The tests in the solar system have established the bending of electromagnetic waves by the distortion of space due to the Sun to within an error of 1 percent of the value predicted by general relativity. The retardation of electromagnetic waves by the field of the Sun has been established to 0.1 percent by ranging to Mars during the Viking program as the line of sight passed close to the Sun. These effects are first order in the field strength and when cast into the formalism of the parametric post-Newtonian (PPN) parameters show that gravity distorts space (i.e., these effects measure the PPN parameter, γ). The advance of the perihelion of Mercury is an effect that depends both on spatial curvature γ and on the time part of the metric to second order in the dimensionless field quantity β. The perihelion advance is in agreement with general relativity to an error of 0.5 percent providing that no corrections have to be applied for a possible mass quadrupole moment of the Sun and that other of the PPN parameters can be assumed to have known values. The perihelion advances of Mars and the asteroid Icarus have also been measured and are in agreement with general relativity to an error of 20 percent.

The gravitational red shift is now confirmed to 0.01 percent uncertainty in the field of the Earth using atomic clock comparisons between a ground-based and a suborbital space platform. The measurements of the red shift can be interpreted as a test of the principle of equivalence and the hypothesis that gravitation is a metric phenomenon. The same term in the PPN expansion that provides the Newtonian limit also is responsible for the red shift in first order of the field strength.

The binary pulsar system is an elegant system in which to test weak-field gravity. The field strengths are comparable to those at the surface of the Sun. More importantly, the period of the motion of one neutron star around the other is a mere 8 h. The periastron advance of the orbit is 4 degrees per year (Mercury's advance is 40 seconds per century), while the relativistic time delay across the orbit is measured in milliseconds rather than hundreds of microseconds as in the solar system. The continuing program to monitor the binary pulsar will be enormously profitable. However, should anomalies be found, one must be sure they are not a property of the timekeeping of the pulsar or an effect of the interaction of the binary system with other stellar systems or components of the interstellar medium.

The binary pulsar system is not a substitute for solar system tests in which all the parameters can be directly measured. The continuation of the precision measurements of planetary ephemerides and the opportunities provided in the space program to improve planetary range data through planetary orbiters and landers leave a rich legacy for experimental tests of weak-field gravity. The value of the data base improves with time because some of the effects due to relativistic corrections—such as determinations of perihelia advances and possible variations in the gravitational constant—grow with time. In addition, many of the Newtonian corrections due to perturbations by other bodies in the solar system become more reliably estimated with time.

The technology has advanced to the point that one can consider carrying out direct tests in a weak field to second order in the field strength. Although it does not follow that the strong-field behavior of relativistic gravitation would be determined by establishing the second-order terms, it must be true that if violations are observed the strong-field solutions would have to be modified. A measurement of the gravitational red shift near the Sun to second order would give an independent estimate of the PPN parameter β. The determination of the gravitational bending of light by the Sun to the second order would yield both new estimates of β and the second-order term in the metric responsible for spatial curvature. Both of these experiments are candidates for space missions.

STRONG FIELD

A profound prediction of general relativity is that regions of space-time exist in which the gravitational field is so strong that they are cut off from our direct observation and that, once entered, there is no escape from them. These regions include sufficient mass, m, in a spatial dimension, r, that GM/Rc^2 approaches 1 on a surface called the horizon that separates the outside and the inside worlds. Observers viewing the formation of a horizon from the outside would see the interior vanish in a finite time (an observer at the horizon of a region containing a solar mass would see the interior vanish in about 10 μs). Once the horizon has formed, phenomena near the horizon appear frozen to a distant observer owing to the large gravitational red shift. To a local observer the horizon behaves as a one-way membrane that permits entry but no exit. Because of this property these regions have been dubbed black holes.

Theoretical work during the past two decades has discovered

that black holes are completely characterized by their mass, electric charge, and angular momentum. All previous history of their constituents is lost during the formation. The black holes may change their charge and angular momentum in interactions with their surroundings, but the intrinsic mass of the black holes never decreases. This is not strictly true, as discovered by Hawking, inasmuch as a black hole has a finite entropy; it will emit thermal radiation by quantum mechanical tunneling at the horizon. Through this process, small black holes will evaporate (and save the second law of thermodynamics). The evaporation time is proportional to the cube of the mass of the hole; a 10^{-18} solar mass hole will evaporate in a time equal to the age of the universe.

Aside from limits imposed by the lifetime due to thermal evaporation, there is no a priori reason intrinsic to gravitational theory to choose masses for black holes. One has to apply astrophysical reasoning to estimate the probability that nature will create black holes and to place limits on their masses. Black holes may have been created in the primeval universe or subsequently in the cores of galaxies by accretion. A strong possibility for creation occurs in the collapse of stars that have spent their nuclear fuel and in which the stellar cores are more massive than can be maintained against gravity by neutron degeneracy pressure. Current wisdom holds that compact objects with masses greater than a few solar masses are black holes.

Black holes are the most efficient converters of rest mass to energy posited in nature. Objects falling into spinning holes could release up to 10 percent of their rest mass into the exterior world. Much of the energy would be released as gravitational radiation due both to the acceleration of the infalling object and the excitation of oscillations of the metric normal modes of the hole. The high conversion efficiency is the main reason why black holes are invoked as the power source for quasars and active galaxies.

A technique to search for black holes is to find dynamical systems in which one has to infer the existence of a dark compact object to explain the motions of surrounding bodies. Searches of this nature have been made in the optical, infrared, and radio wavelengths using high spatial resolution coupled to spectroscopy. The "great observatories" proposed by the space program will play an important role in this search. The most promising black hole candidates have been uncovered in x-ray astronomy. The Cygnus X1 source is part of a binary system in which one component is a visible star that appears to be feeding a compact object of greater than 3 solar masses. The x-ray emitter must be of small size since the source is seen to fluctuate on time scales as short as milliseconds. The Advanced X-ray Astrophysics Facility (AXAF) will uncover more such systems, and large area x-ray detectors will be useful in measuring the fast time fluctuations. One of the more dramatic prospects for gravitational wave astronomy is the detection of the metric ringing of a black hole after excitation by the infall of matter. The signatures of the signals are directly calculable; the oscillation periods are proportional to the mass of the hole. A 10-solar-mass black hole will produce a pulse with strong 1-kHz components and is detectable using terrestrial instruments. The search for signals from holes with mass larger than 10^4 solar masses is best carried out by space-borne gravitational wave detectors.

NONSTATIC FIELDS

When the sources of a gravitational field are in motion, new phenomena occur in relativistic gravitation that have no analogs in Newtonian gravity but can be thought of as having closer analogies to electromagnetism. In particular, magnetic gravitational effects and gravitational radiation are two phenomena that are amenable to measurement.

The weak-field solution of the gravitational metric outside of a spinning sphere was presented by Thirring and Lense in 1918. The spinning sphere, in addition to producing spatial curvature, causes the inertial frame in its vicinity to be dragged along with the rotation by an amount of the order of the gravitational field strength multiplied by the rotation rate. In an analogy with electromagnetism, the spinning sphere produces an "electric" gravitational interaction due to its mass, represented by the pure time part of the metric, which yields the ordinary Newtonian potential. It also generates a "magnetic" gravitational interaction due to the mass currents that produce a gravitational vector potential represented by the crossed time-space terms in the metric. The gravitational magnetic terms interact with mass currents much as magnetic fields interact with ordinary currents. At the time, Thirring and Lense looked at the possibilities of determining "magnetic" gravitational effects on the satellites of Jupiter, but concluded that the effects on the orbits would be too small to measure. In the early

1960s, L. Schiff analyzed the motion of a gyro (a gravitational magnetic moment) in the field of the spinning Earth and realized that the precession of the gyro could be large enough to measure. An attempt to do this is embodied in the experiment known as Gravity Probe B, which will be discussed below. The experiment will be a fundamental measurement of a new phenomenon in gravitation, which is expected to play an important role in some astrophysical problems. Present-day models of double-lobe radio sources invoke a spinning black hole that powers the radio source and gives directional stability to the plasma jets connecting the lobes. Frame dragging at the hole may be the mechanism providing the power.

An interesting sidelight is Thirring's analysis of the weak-field solutions inside a spinning shell of mass in which Coriolis and centrifugal forces are generated by the spinning shell. He alluded to the possibility that Mach's principle could be directly incorporated into relativistic gravitation if one were able to solve the cosmological boundary value problem.

In 1918, Einstein established that time-dependent free field solutions to the gravitational field equations should exist. These are gravitational waves traveling at the speed of light that distort space-time transverse to their propagation direction. The plane wave solutions are visualized as strains in space that contract space in one dimension and expand it in an orthogonal dimension, both of which are transverse to the wave propagation. Two polarizations are possible, differing only in that their strain axes are rotated by 45 degrees. The waves carry energy away from their sources and, just as in electromagnetism, exert damping on the radiating systems. Strong evidence for gravitational radiation damping has now been observed in the binary neutron star system PSR 1913+16. The sources of the gravitational radiation are accelerating masses. Since there is only one sign of the mass and since the center of mass of an isolated radiating system must remain fixed, the lowest order of the radiation is derived from the time-varying quadrupole moment of the source. In a quantum mechanical description of general relativity, which is a tensor theory of rank 2, the field particle—the gravitation—will have a spin of 2.

The direct detection of a gravitational wave may be accomplished by several methods. One method is the mensuration of space between free bodies that "ride" the wave. This is done by measuring changes in the time it takes electromagnetic waves to travel between a configuration of free masses. An alternative method is to measure the time-dependent distortions of extended bodies under the influence of the gravitational wave reexpressed in terms of "tidal forces." Both techniques are being actively pursued in current research and will be described more extensively later.

The direct experimental confirmation of the wave solutions by laboratory experiments using a terrestrial source and receiver, in analogy to the Hertz experiment for electromagnetic waves, appears to be impossible with present technology or that of the forseeable future. As a consequence, the direct detection of gravitational waves has become an astrophysical problem, for it is in this realm that we expect to observe phenomena in which the coherent motions of large masses at high velocity will make measurable gravitational wave strains at the Earth. The coupling of astrophysics to the fundamental study of the gravitational interaction makes the search for gravitational radiation a particularly exciting field. Not only will we test gravitational theory by investigating the wave solutions and gather evidence of the behavior of strong-field gravity at the sources, but we expect to gain a new view of the universe through the gravitational wave window. The space program offers a particularly important opportunity for this research, as it enables observation of gravitational radiation in the low-frequency bands, which may be inaccessible or unobservable from the ground (see Chapter 3).

COSMOLOGY

In principle, the large-scale properties of the universe are described by a solution of the field equations of relativistic gravity coupled to the equation of state for matter and a set of initial conditions. At present, the grand synthesis of cosmological theory is still only a dream, but the essential ingredients are expected to involve the unification of the forces known in nature and a viable quantization of gravitation. The realization that understanding the origin and evolution of the universe will require a deeper understanding of all of physics has long been suspected, but the observational and theoretical discoveries of the past two decades have given the notion real substance.

Cosmological studies have always been a singular branch of science. There is only one universe to observe, and so the standard techniques of classification and comparative studies, familiar and so useful in astronomy, do not apply. We must perform our observations from within; there is no way to step outside and observe the entire system. Cosmology, in common with the rest of astronomy, does not permit us to alter conditions in the system and then observe the effect of our tinkering, a practice that is so important in an experimental science. For all of these reasons, cosmological research is more dependent on the interaction of theoretical development with observation than most other branches of natural science. The only hard discriminants indicating that one is on the right track come through the consistency of one cosmological observation as related to another through a theoretical framework.

The explosive origin of the universe is strongly suggested by the red shift distance relation first discovered in the 1920s. It is also consistent with the discovery of the 3K cosmic background radiation in the 1960s. Although the "big bang" model is not unique in relating the known cosmological observables, it is at present the model making the least ad hoc assumptions and serves as a useful and possibly correct framework to describe cosmological studies. An overview of the issues in cosmology is conveniently given by the epochs in this model.

In the modern epoch the universe is endowed with stars and galaxies, the arena of astronomy. The explored part of this epoch extends back in time to when the universe was approximately half its present size and twice as hot. The evolution and geometry of the universe in this epoch are expected to be dominated by the matter density. The research associated with this epoch is crudely characterized by cartography and by taking inventory of the mass. One aim is to relate the global geometry to the matter distribution through the Einstein field equations. A longstanding hope is to measure the geometry with sufficient freedom from systematic errors such as evolutionary effects to determine whether the universe is opened or closed. The observational problem is to establish the cosmic distance scale by coupling red shift measurements with reliable absolute sizes or luminosities of astronomical objects. Advances in detector technology and the planned large-aperture, ground-based and space-borne observatories are expected to make significant progress in this program by observing larger and fainter parts of the universe.

Associated with determining the global geometry of the universe is the exploration of its large-scale structure. The clustering of matter and the fixing of local velocity fields may be remnants of primeval large-scale density fluctuations, if these phenomena were not extensively modified by dynamical processes occurring at later times.

The converse problem of determining the mass in the modern universe seems more difficult. A longstanding observation is that the luminous matter accounts for less than 10^{-2} of the mass required to close the universe. Dynamical methods of measuring mass indicate much larger values. The rotation curves of individual galaxies imply that at least twice as much mass is invisible (dark mass). The dynamics of galactic clusters imply that 30 to 100 times the amount of luminous mass is dark. The nature of the dark matter in the universe is a major puzzle. Dark matter candidates must be consistent with other cosmological observables, in particular the measurements of the 3K cosmic background and the measured isotopic abundances of light elements. Space-borne astronomy may provide the critical clue by opening up the entire electromagnetic spectrum.

A poorly understood epoch is the transition from the time when the 3K background was last scattered by matter to the formation of the modern universe. The expectation is that during this time the first stars or galaxies were formed out of density fluctuations that already had to exist in earlier epochs. An observable remnant of this period could be the radiation emitted by the release of energy due to nuclear burning. Although it is not expected that the 3K background is strongly perturbed by the plasma in this period, it is possible that fine-scale anisotropies are washed out by Thomson scattering, thereby invalidating conclusions now being drawn from the isotropy of the 3K background. Space-borne observations, especially in the infrared, may shed light on this epoch, if the emission from foreground sources is not overwhelming.

The 3K background was last scattered by matter at the time when the primeval plasma condensed into atoms, at a temperature of 10^4 degrees when the universe was 10^{-4} of its present size. Epochs prior to this are not observable directly by electromagnetic astronomies. The 3K background is therefore one of the few relics conveying information on earlier epochs of cosmic evolution. The radiation has been measured over three decades of the electromagnetic spectrum, including both the Rayleigh and the Jeans and Wien parts of a thermal source. The data, of varying quality, are represented by a single temperature to approximately 10 percent uncertainty. The precision of the measurements is not sufficient to set meaningful limits on likely processes in the early universe that might not have equilibrated. Conventional wisdom holds that the spectrum will not be easy to perturb because the heat capacity of the radiation so overwhelms that of matter in the early universe. The photon to baryon ratio is of the order of 10^9.

Measurements of the angular distribution of the 3K have presented a challenge to cosmology. A large-scale anisotropy interpreted as being due to the motion of the observer relative to the last scatterers of the radiation has been measured. The velocity of our galaxy determined from this anisotropy is larger than expected. More important, it has set a benchmark for mapping velocity fields in the modern universe. No other anisotropy on angular scales extending from 90 degrees to a few minutes of arc has been measured to a level of 10^{-4}. The most stringent upper limit is 2×10^{-5} at an angular scale close to 3 minutes.

The absence of anisotropy raises at least two questions. The first, more philosophical than the second, is how might the universe be created so that large-scale regions (1 degree or larger) that could not have been causally connected throughout the expansion would have the same temperature? An answer is proposed by the inflationary universe model. The second question is how did galaxies form in the short time after decoupling of the radiation from the matter? The initial expectation was that galaxy formation would be the end result of the growth of adiabatic density fluctuations in the primeval plasma. The fluctuations would leave imprints of small-scale anisotropy in the 3K background at a level of 10^{-4} or larger. The high degree of isotropy has led to speculations that the primary density fluctuations are isothermal and due to condensations of nonbaryonic unknown particles. The new particles, if long-lived, could be candidates for the invisible mass, but must be tailored to have properties that would not upset the isotopic abundances determined in the earlier epoch of primeval nucleosynthesis.

The space program offers important opportunities for the observation of the 3K background. Measurements of the spectrum and large-scale anisotropy will be carried out by the Cosmic Background Explorer (COBE) satellite. Should the results of this mission indicate that foreground emission does not compromise further observations at higher sensitivity, such observations should be considered for future missions. A new possibility for 3K observations is the use of space-borne millimeter dishes to measure small-scale anisotropy. The millimeter-wave region lies at frequencies above the contamination by galactic synchrotron radiation and below the emission of interstellar dust but is perturbed by atmospheric emission in ground-based telescopes.

A constraint is imposed on cosmic models by the epoch of nucleo-synthesis. As the universe cooled, it spent a few minutes at a temperature where fusion of baryons was possible. The primordial abundance ratios of the light elements was established at this time. The relative abundances are calculated by nuclear reaction theory with the cosmic temperature, the baryon density, and the expansion time as input variables. The abundances of H_2, He_3, He_4, and Li_7 measured in the interstellar medium and in the spectra of old stars is in remarkable agreement with the cosmic model, providing that the photon to baryon ratio is not smaller than a few times 10^9. Of the more abundant isotopes, the primordial abundance of H_2 is particularly sensitive to the mass density, decreasing with increasing mass density. On the other hand, He_4 has little sensitivity. Should the measurements of primeval isotopic abundances hold as more regions of the universe are observed, it will place strong constraints on the constituents of hidden matter, indicating that no more than 10 percent of the mass in the universe is baryonic.

The epoch of nucleosynthesis is the earliest from which any hard evidence of cosmic evolution is available at present. Two other relic backgrounds have been hypothesized. The first is a thermal bath of neutrinos, now near 1K, which decoupled from matter at a temperature of $10^{11}K$ when the muons annihilated. The second is a stochastic background of gravitational radiation that might have originated during the Planck epoch when the universe was opaque to gravitational radiation.

The realization that the explosive universe may have experienced temperatures as high as $10^{61}K$ (10^{57} eV) and mass densities as high as 10^{93} g/cm^3 in the Planck era, when a still unformulated quantum theory of gravitation dominated the microphysics, and that the universe subsequently passed through all imagined and known domains of high-energy physics, has led to a synthesis of theoretical cosmology and quantum field theory. Several new ideas have emerged from this synthesis. One is a speculation that at sufficiently high energies (10^{24} eV) all the interactions in nature (but not gravity) have comparable strengths. It has been suggested that at these energies, through the mediation of a new boson that couples the quark and lepton fields (responsible for proton decay), it is possible to violate baryon conservation. If this is true, it could offer an explanation for the antimatter/matter asymmetry seen in the universe and furthermore would account for the development of the baryon to photon ratio we observe, rather than impose it as an initial condition.

Another idea, which grew out of theoretical work associated with gauge fields, is that of the inflationary universe. In this model of the very early universe there is no matter or radiation density at the beginning, but a large fixed energy stored in a symmetric state of the vacuum. The cosmological metric evolves exponentially, just as the de Sitter (empty) universe with a positive cosmological constant, until the vacuum has undergone a transition to a lower energy state (assumed to be zero) with broken symmetry. At this point the universe has expanded by a factor of 10^{28}, and the latent energy of the initial vacuum state has been transformed into the matter and radiation fields that from then on determine the more conventional evolution of the universe. The inflationary universe model provides an explanation of the large-scale isotropy of the 3K background by having the entire universe expand rapidly from a single causally connected region of the vacuum. Furthermore, the model predicts that the mass density should be exactly equal to the closure density.

The coupling of field theory with cosmology is only beginning. The unification of gravitation with the other fields in nature will be a key ingredient in theories of the very early universe.

It is too early to tell if the space program will play a direct role in this aspect of cosmological research. One could imagine that some compelling candidate particle that might only be detectable above the Earth's atmosphere will be proposed as a constituent of the dark matter. At present this does not seem to be the case.

3
Gravitational Wave Astronomy

A major challenge and opportunity in relativistic gravitation is the direct detection of gravitational radiation from astrophysical sources. The expectation is that technology now under development will within the next 10 years detect gravitational radiation or, if nature is insufficiently kind, that we will at least be able to set interesting astrophysical limits on the gravitational waves incident on the Earth. These terrestrial observations will be made at gravitational wave frequencies above 10 Hz. The exploration of the potentially rich low-frequency spectrum of gravitational waves can only be carried out from space. The development of low-frequency gravitational wave detectors and ultimately a space gravitational wave observatory is a major new direction proposed in this study for the space program.

The observation of gravitational radiation from astrophysical sources has several features. First, the direct detection of the waves will serve to test relativistic gravitation by measuring the propagation speed and polarization states of the waves. The sources of the waves will most likely be regions in which the gravitational field strength is large. Thus, the detection of signals from these regions will serve to test gravitation in the strong-field, high-velocity limit. Second, gravitational radiation is very weakly coupled to matter and will not scatter even in the strongest sources. Observations of phenomena deep in the interior of regions normally obscured in the electromagnetic astronomies will thus become accessible. The processes involved in stellar collapse and the primeval cosmic kernel are two well-known examples.

Estimates of the gravitational wave flux incident on the Earth from astrophysical sources have been divided into three categories: sources of gravitational wave impulses or bursts, periodic sources that may produce continuous gravitational wave trains, and, finally, sources of a gravitational wave background noise. The categories are related to different techniques for detection. The estimates themselves are of varying quality. In some cases, such as the radiation from ordinary binary stellar systems, the estimate is reliable. Should no radiation be observed, it would indicate a failure of the theory. For other sources, such as supernova explosions, the occurrence of the phenomena is well established but our ignorance of the physical processes involved leave the amplitude of the waves uncertain. Yet another class consists of posited sources whose number is unknown, but for which the amplitude of the waves is calculable; black holes are the best example of this class. Prediction in astrophysics is always a hazardous exercise, especially when one opens a new field or when a profound change in sensitivity makes new observations possible. The theoretical predictions of what might be discovered at the birth of x-ray astronomy certainly did not anticipate the enormous diversity of the phenomena that subsequently were uncovered.

Supernova explosions are known to occur in our galaxy at a rate of between 1 and 10 per century. The collapse of the core of a supernova to a neutron star or black hole could be accompanied by the release of gravitational radiation if the collapse is not spherically symmetric. The time scales of the collapse during which gravitational radiation would be emitted lie between 1 and 10 ms, but the fraction of the explosion energy going into gravitational radiation is uncertain. A supernova at the center of our galaxy that releases 1 percent of its energy into gravitational radiation would produce a wave amplitude at the Earth having a strain of 10^{-18}. Present-day gravitational antennas would detect such a pulse. In part, the goal of ground-based efforts to detect gravitational radiation has been set by the search for supernova events to achieve a strain sensitivity of 10^{-21}, which would observe these events to a distance of the Virgo cluster of galaxies at an event rate of 1 to 10 per year.

Pulsars are periodic sources whose number of occurrences is known but whose amplitude of gravitational radiation is uncertain. The amplitude depends on the mass eccentricity of the spinning star. Should the eccentricity be entirely due to distortion of the star by the magnetic fields trapped in the star during the collapse from an ordinary star, the wave amplitudes from pulsars in our own galaxy would give strains of 10^{-32} to 10^{-33}, much too small to measure. However, the pulsar could have an intrinsic mass eccentricity, which, if it was as large as 10^{-5}, would produce a measurable strain of 10^{-26}. Earth-based detectors with wave frequencies above 10 Hz are now being planned with such sensitivities. Many of the pulsars have frequencies around 1 Hz and thus would be candidates for space antennas.

Ordinary binary stellar systems abound in the galaxy; approximately 1 percent of all stars are members of binary systems. The closer and fast ordinary binaries are sources of gravitational radiation at strain levels of 10^{-21} with periods ranging between 1 to 10 h. These are clear candidates for detection by space antennas. In fact, there may be so many of them that they could constitute an unresolved stochastic background of gravitational waves. A subclass of binary stars includes the double neutron star systems such as PSR 1913+16. These are particularly interesting for both space- and ground-based gravitational antennas. PSR 1913+16 now radiates at submultiples of 8 h with strain amplitudes around 10^{-23}, which would be detectable by a space antenna. After about one million years, as this system loses energy through gravitational radiation, it will produce a gravitational wave chirp that will be detectable by ground-based antennas. The system will then spend about 1 year near a period of 10 s and ultimately come to an abrupt end in 1 ms as the two stars collide. The gravitational wave strain multiplied by the square root of the number of cycles lies in the vicinity of 10^{-18}. Three binary neutron star systems from a total population of several hundred pulsars are known to exist in our galaxy. Extrapolating to the rest of the universe, we could expect to detect an event of this type every few hours in an antenna with a strain sensitivity of 10^{-22} to 10^{-23}.

Binary systems composed of double white dwarf stars are fairly certain to exist. These systems radiate at periods of 1000 to 100 s with strain amplitudes in the region 10^{-20} to 10^{-22}. Not observable by ground-based antennas, they fit well into the best performance region of a projected space antenna.

There are a host of more speculative sources. The gravitational radiation from the collision of a black hole with another black hole or other compact object as well as the radiation emitted in the formation of a black hole is well studied. The radiation originates both from the acceleration of the masses and from the excitation of ringing in the normal modes of the metric solution around the black hole. The gravitational wave bursts from such events have large strain amplitudes with frequency components that vary as the inverse of the black hole masses. The formation of a 10-solar-mass black hole at a distance of 100 megaparsecs (Mpc) could produce a strain pulse of 10^{-21} lasting 1 ms. Should black holes form binary systems, the orbital decay of a 10-solar-mass black hole binary system anywhere in the universe could be observed with an antenna having a strain sensitivity of 10^{-22}. The space antennas are well suited to measure the radiation from massive black holes. The formation of a 10^7-solar-mass black hole anywhere in the universe would produce a strain at periods of several hours of 10^{-16} or larger.

One of the most interesting speculative sources of gravitational radiation under current theoretical consideration is the radiation suffusing the universe that may have originated in the universe's earliest epoch. Should present thinking be correct, it is possible that quantum gravitational wave fluctuations during the Planck epoch were amplified in the subsequent universal expansion. The radiation would appear as a gravitational wave background noise. The spectrum of the radiation is not well understood; however, it is believed to contain less than 10^{-4} of the energy density required to spatially close the universe. The search for cosmic background of gravitational radiation is a prime motivation for both space- and ground-based gravitational wave antennas.

The sensitivity of gravitational wave observations on the ground is advancing rapidly. Acoustic bar detectors at cryogenic temperature using low noise position transducers are now able to search for gravitational wave bursts in the kilohertz band with a strain sensitivity of 10^{-18}. Several detectors are now operating in coincidence, and results of the searches at this level of sensitivity should be available in 1986. The acoustic detectors will continue to improve but must be able to circumvent the "naive" quantum limit, which will set in at strain sensitivities between 10^{-20} and

10^{-21} depending on the frequency of observation. Acoustic detectors will be constructed at lower frequency and as the transducer technology improves could have bandwidths of around 10 percent.

The other ground-based technique for detecting gravitational radiation utilizes laser interferometer systems that measure the separation of a configuration of free masses. This technique is also a precursor for the most promising space antennas. At present the largest of these systems are 30 to 40 m long. Using light powers of several hundred milliwatts, they can attain strain sensitivities of 10^{-17} at 1 kHz. The promise for greatly enhanced sensitivity lies in constructing these systems with 100 times larger length and increased position sensitivity by increasing the light power modulated by the interferometer. The systems are inherently broadband, and as the techniques for reducing the stochastic forces on the masses improve, they are expected to perform at a pulse strain sensitivity of better than 10^{-22} between 10 Hz and 1 kHz. The sensitivity for periodic sources is expected to be less than 10^{-26} for integration times of a month. The present plan is to construct two 4-km-long antennas in the United States, and there are plans in Great Britain, Germany, and France to construct antennas of comparable length. These antennas will be operated as a network to determine the position of gravitational wave sources in the sky. These systems will be limited by ground noise and gravity gradient noise to operate above 10 Hz. The exploration of the gravity wave flux at longer periods is clearly the domain of space research, where longer baselines are possible and smaller low-frequency stochastic forces will be encountered.

4
Current Space Research in Gravitation

At present, the U.S. space program is involved in gravitational research in several areas. The lunar ranging program and the continuing program to establish the motions of solar system bodies constitute an important ongoing research area. These are described in this chapter. The development of the magnetic gravity gyro experiment (the major gravitational program at NASA), the development of the cryogenic principle of equivalence experiment (supported by PACE and NSF), and the search for long-period gravitational radiation by tracking Galileo and Ulysses are described in subsequent chapters.

LUNAR RANGING

The optical retroreflector packages placed on the lunar surface by the Apollo 11, 14, and 15 missions and by Luna 21 make possible highly accurate laser distance measurements to the Moon. The large majority of the data through 1982 was acquired by the McDonald Observatory in Texas. The accuracy is typically 10 to 15 cm.

Recently, three other lunar ranging stations have begun regular observations. These stations are in France, Australia, and Hawaii. Three of the four stations have 1.5-m-diameter receiving telescopes, and all four are expected to have sub-nanosecond pulse-length lasers soon. A precision of 1 to 2 cm for roughly 15 min of observing time has been demonstrated, and routine performance with similar accuracy is expected.

Nordtvedt pointed out in 1968 that lunar range data would provide a sensitive test of whether the gravitational self-energy of the Earth obeys the principle of equivalence. From one point of view, this test gives a check on one of the fundamental assumptions of general relativity. From a different viewpoint, the lunar distance provides the best test at present of the superposition law for gravitation, and thus can be regarded as a fifth test of general relativity. Within the framework of conservative theories without a preferred frame, the size of the effect is given by

$$\Delta d = (4\beta - \gamma - 3)d_N \cos D,$$

where $d_N \sim 10$ m and D is the difference in mean longitudes of the Moon and Sun. When combined with the value for γ from the Viking time-delay measurements, the present lunar ranging results show $\beta = 1$, as predicted by general relativity, with an uncertainty of 0.004.

The lunar ranging measurements are likely to continue to play a substantial role in solar system tests of gravitational theory in the future. With the current improvements in measurement accuracy, an uncertainty of less than 0.001 for β is expected. A determination of geodetic precession for the lunar orbit should be possible soon, and will improve as a longer span of accurate data is obtained. In addition, lunar range data will give an independent check on the constancy of the Newtonian gravitational constant with an expected accuracy of better than one part in 10^{11} per year. Lunar range data also complement planetary distance measurements by helping to tie down some of the classical parameters needed in order to test relativistic predictions.

ANALYSIS OF PLANETARY AND LUNAR MOTION

An important phase of present research in gravitational physics is the analysis of spacecraft tracking data, lunar laser ranging data, and planetary radar measurements in order to determine the dynamics of the solar system. The motions of the inner planets and the Moon provide our best tests of several gravitational phenomena; measurements of the relativistic time delay for electromagnetic signals passing near the Sun also are of major importance. Striking progress has been made in the last few years by analyzing tracking data for the Viking Lander and Orbiter spacecraft, in combination with tracking data for other spacecraft, lunar ranging data, planetary radar data, and other solar system information.

One advantage of combined solutions using all available data is that the time base for the observations extends over a longer period than for individual spacecraft missions. Even though the 1971-1972 tracking data for the Mariner 9 Mars Orbiter gave less accurate Earth-Mars distances than the Viking mission provided, the inclusion of these earlier Mariner 9 observations yields constraints on the orbital motions over a substantially longer time. Since nongravitational forces on the planets and the Moon are negligible, the integrated effects of non-Newtonian effects on orbital motions can be determined more accurately with the extended data sets. Although optical observations of the planets are less accurate than spacecraft tracking or radar measurements, they help to determine some orbit parameters to which distance measurements are less sensitive. Since the orbit of the Earth is common to all of the data types, improvements in it from one type of observation help to increase the strength of the other data types. Thus, all of the accurate data on the motions of the inner planets and the Moon need to be analyzed jointly as an integrated test of the extent to which solar system dynamics obey our current understanding of the laws of gravitation.

It is not clear whether new tracking data for planetary orbiters or landers that is useful for gravitational physics will be obtained in the next decade. However, some new radar distance measurements to Mercury are being made, and they could provide substantial improvements in our knowledge of Mercury's perihelion precession. Lunar laser range data of improved accuracy and from a number of stations are now being obtained, as discussed previously. In addition, more refined analyses of the Viking tracking data and of the effects of the asteroids on the motion of Mars are very much needed. In view of the complexity of the solar system and the great difficulty of modeling all of the interactions between the different bodies well enough to give their positions with accuracies on the order of one part in 10^{12}, it is essential that the work be carried out by at least two independent groups. Such major analysis efforts currently are being carried out by the collaboration between MIT, Harvard University, and the Smithsonian Astrophysical Observatory and by the Jet Propulsion Laboratory of Cal Tech. The Task Group on Fundamental Physics and Chemistry believes that strong continued support is needed for such independent but mutually supportive efforts in order to achieve continued progress on solar system dynamics tests of gravitational physics during the next decade. More intensive programs of radar distance measurements to Mercury that make use of repeatedly observed "closure points" to reduce uncertainties from planetary topography would be valuable in providing improved tests of gravitation, as would increased accuracy for multiple-station lunar laser range measurements.

5
Expected Research Prior to 1995

SHUTTLE TEST AND FLIGHT OF GRAVITY PROBE B (GPB)

Gravity Probe B (GPB) is a gyro test of general relativity made with four cryogenic gyroscopes whose spin axes are compared with the position of a star to 10^{-3} sec of arc in a zero-g satellite for a period of at least one year (see Figure 5.1a). For the first time the experiment tests the dragging of inertial frames (magnetic gravity) in general relativity due to the rotating Earth and the geodetic precession due to the motion of the gyros around the Earth (see Figure 5.1b). It is planned that the experimental package will be tested on a Shuttle flight in 1989, followed by a free flyer zero-g experiment to perform the actual relativity experiment in 1991 or 1992.

The Shuttle test is an important step in the GPB program. It will test the completely integrated package consisting of the instrument, dewar, and electronics. The Shuttle test will provide some information on gyro drift at reduced g but will not be able to establish the low drift rates of the gyros required to carry out the actual experiment.

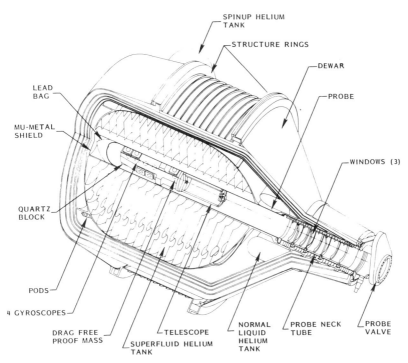

FIGURE 5.1a Gravity Probe B (GPB) experiment module. SOURCE: Reprinted from D. Bardas et al., "Hardware Development for Gravity Probe B," in *Proceedings of SPIE, the International Society for Optical Engineering*, volume 619, page 33, 1986.

SHUTTLE FLIGHT OF A CRYOGENIC PRINCIPLE OF EQUIVALENCE EXPERIMENT

The principle of equivalence requires that the ratio of the inertial to gravitational mass of all bodies be the same. An experiment done in the mid-1960s has established that this ratio is the same for gold and aluminum to one part in 10^{11}. New experiments to set better limits are in progress at Joint Institute for Laboratory Astrophysics (JILA) and at Stanford University. The Stanford experiment is being designed to be placed in space.

A superconducting equivalence experiment has been designed with the potential of testing the equivalence of gravitational and inertial mass to a projected sensitivity of one part in 10^{15} on a Shuttle flight and one part in 10^{18} on a zero-g free flyer. At

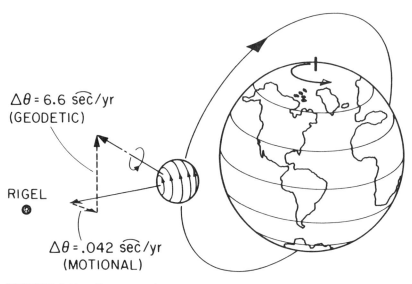

FIGURE 5.1b Gyro experiment orbit. Relativistic effects as seen in gyroscope with spin vector oriented as shown and lying parallel to the line of sight to Rigel. SOURCE: Reprinted from W. Fairbank, *Near Zero: New Frontiers of Physics*, W. H. Freeman, New York, 1987.

this level the experiment tests with an improvement of 3 and 6 orders of magnitude, respectively, the foundation on which the geometrization of space-time in Einstein's general theory of relativity is based. A violation at any level would pose a problem for general relativity. A new long-range force coupled to baryon number or further peculiarities in the weak interactions are posited sources of such a violation.

Conceptually, the experiment is similar to Galileo's purported experiment of dropping different weights from the Leaning Tower of Pisa, except that instead of falling a few tens of meters, the objects fall all the way around the Earth. Two concentric test masses, a solid rod and a hollow cylinder, are constrained by magnetic bearings so that they are free to move only along a common axis (see Figure 5.2). As they orbit the Earth with fixed orientation in inertial space, they are subjected simultaneously to the centrifugal acceleration of the orbital motion and the gravitational attraction of the Earth. If the ratios of gravitational to inertial

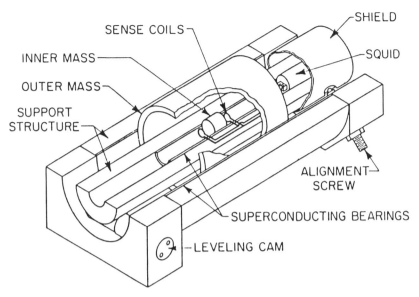

FIGURE 5.2 Equivalence principle accelerometer; cryogenic Eotvos experiment. SOURCE: Reprinted from C.W.F. Everitt and Paul W. Worden, Jr., *A Preliminary Study of a Cryogenic Equivalence Principle Experiment on Shuttle*, page 20, 1985.

mass of the test bodies are different, there will be a periodic differential acceleration between them, which will show up as a relative displacement at orbital period along their axis.

On the Shuttle the experiment is limited to one part in 10^{15} by the changing gravitational gradients due to motion of mass on the Shuttle. On a specially designed drag-free satellite the potential accuracy of the experiment is one part in 10^{18}. One feature of the experiment is that the perturbing effects due to gravitational field gradients can be reduced to negligible levels by making the center of mass of the two test masses coincident. This is done by sensing the differential displacement signals at twice the orbital frequency and using the signature in a null servo system.

MICROWAVE RANGING TO THE MARS OBSERVER SPACECRAFT

The planned orbit for the Mars Observer spacecraft is sun-synchronous and nearly circular, with 93-degree inclination and 361-km altitude. The telecommunications system will permit

X-band Doppler measurements, but no ranging signals or dual-frequency capability is planned. Thus, unless at least ranging signals are added, no information useful for gravitational physics can be obtained.

With dual-frequency capability and ranging signals added, the Mars Observer mission could improve on present knowledge of solar system dynamics. The first requirement would be to determine the gravity field for Mars accurately so that the Earth-spacecraft distance could be converted to the Earth-Mars center-of-mass separation. Despite the low altitude, and the corresponding high-degree and high-order gravity field solution required, valuable results could be obtained if the range and Doppler measurements were sufficiently accurate. There also may be an opportunity for accurate lower degree and lower order gravity field solutions and thus improved Earth-Mars distance determinations when the spacecraft orbital altitude is raised at the end of the planetary observation period. This would be possible only if the operation of the spacecraft altitude control system and tracking system could be continued.

New Earth-Mars distance measurements starting in 1991 with the arrival of the Observer spacecraft at Mars could be combined with the Viking Lander tracking data during the period 1976 to 1982 to make the duration of the observations 3 times longer. This would result in better knowledge of the orbits of Mars and Earth, as well as improved masses and densities for a number of the asteroids that significantly perturb the motion of Mars. For testing gravitational theory, the accuracy of the precession of perihelion for Mars and the limit on the possible rate of change of the gravitational constant would be improved significantly. The greatly improved gravity field for Mars also would be of high value for planetary studies. However, in view of the present lack of plans for ranging to the Mars Observer spacecraft, we cannot expect information of the kind discussed above to be obtained during the next decade. The task group does not know when another comparable opportunity may occur.

X-RAY TIMING EXPERIMENT AND A MEDIUM-AREA FAST X-RAY DETECTOR ON THE SHUTTLE

X-ray astronomy is closely linked to gravitational physics because of the fact that most bright x-ray sources are identified either with compact objects such as neutron stars, black holes, and active galactic nuclei or with large clusters of galaxies in which missing mass influences the spatial extent and temperature of the diffuse x-ray emitting gas. X-ray emission in compact objects is produced deep in the potential wells—near or on the surfaces of the neutron stars and near the Schwarzschild radius in the black holes. Effects such as gravitational radiation, the stability of orbits near the Schwarzschild radius, gravitational red shifting, and gravitational lensing become important to understanding how x-ray emission is produced, and the x-ray observations can be utilized to test basic physical principles.

The x-ray sources available for such purposes are very bright. Advanced experiments can be undertaken with this abundant flux by building instruments with appropriate performance characteristics. X-ray sources have other desirable properties; for example, the spherical neutron stars should be gravitational lenses of high optical quality, and their symmetry simplifies lensing calculations. Black holes are thought to be even more perfectly symmetrical than neutron stars, with the metric near such objects being well-specified when given only the mass, angular momentum, and net charge. Accreting material probes that metric. In general, timing and spectral measurements are expected to provide the best probes, provided that there is sufficient sensitivity to observe at the dynamical time scales of the sources—often milliseconds or shorter.

The use of accreting neutron stars to study gravitational radiation instabilities appears to be a field particularly ripe for experimental development. Several theoretical lines of argument lead to the expectation that neutron stars accreting from binary companions should be able to spin up to angular frequencies in which they are subject to relativistic instabilities, with the angular momentum supplied from the accreting disk continually radiated away in gravitational waves. Such objects would be pulsars in both gravitational waves and x-rays. Continuous-wave signals in both types of radiation would be precisely phase-locked with one another. Detection of the x-ray signal would permit a gravity wave antenna to be constructed so as to be optimally tuned for direct detection of the gravity wave flux.

It has become apparent that many extremely fast processes occur in compact x-ray sources, but progress in understanding has been limited by data quality. Larger x-ray detector collecting areas and the capability to deal with very high data rates are needed in order to make timing observations of sufficient detail to advance our understanding of the physics of compact, highly relativistic objects.

A desirable program to implement before 1995 would include the Shuttle flight of a proportional counter array of a few square meters of collecting area with thin windows to allow detection of x-rays from 0.25 keV to 50 keV. The data bit rate may be as large as 10 Mbps. Later, on the Space Station, a possible experiment would entail the construction of a large (100-m^2 effective collecting area) x-ray detector array devoted to fast-timing and time-resolved broadband spectral studies. The Shuttle instrument development can be used to engineer a proportional counter module that could be replicated inexpensively for the large array on the Space Station.

A Shuttle instrument for timing observations of the brightest compact x-ray sources offers the exciting possibility of discovering new phenomena in accreting binary systems containing a neutron star. For example, there are strong reasons to believe that in such a system a neutron star with a weak magnetic field can be spun up by accretion to periods so short that the star is subject to general relativistic instabilities. The nonaxisymmetry resulting from this instability can produce coherent gravitational waves and modulation of the x-ray flux generated by accretion. If even one instance of this phenomenon is found, it will have profound consequences for the gravity wave detection effort. A detection in x-rays wavelengths would allow a coincident, phase-sensitive search with ground-based gravity wave detectors now under development.

A Shuttle flight of a medium-area instrument would also provide a much improved capability to study millisecond x-ray burst activity from the inner part of the accretion disk around massive objects such as black holes. These bursts provide a probe of the metric very near the innermost stable orbit of a black hole.

SPACECRAFT OBSERVATIONS OF LONG-PERIOD GRAVITATIONAL WAVES

The gravitational wave spectrum at periods from a few hours to several minutes may be explored by observing the motions of interplanetary spacecraft. The technique is most sensitive to gravitational wave bursts, with periods shorter than the propagation time between the Earth and the spacecraft. The gravitational wave burst is seen in the Doppler data as a dual pulse with a time difference determined by when the gravitational wave hits the Earth and the spacecraft. The technique is not limited to bursts and could be used in a search for periodic sources. With several spacecraft operating simultaneously, it is possible to carry out coincident searches as well as to observe a stochastic background of gravitational radiation.

Searches for long-period gravitational waves will be made by the Galileo and Ulysses missions. The launch dates for these missions are now in doubt, but simultaneous observations will still be carried out when possible. The root-mean-square strain sensitivity in these searches is anticipated to be 3×10^{-15}, with the noise budget determined almost equally by uncertainties in the electronics, the transmission by the troposphere, and the fluctuations in the interplanetary plasma. In order to measure the fluctuations in the column density of the interplanetary plasma, dual-frequency transmission at both S- and X-band from the spacecraft to the Earth is incorporated in both Galileo and Ulysses. Galileo has the option of S- or X-band transmission from the Earth to the spacecraft, while Ulysses can use only S-band. Similar experiments have been proposed for the Mars Observer mission and the Comet Rendezvous Asteroid Flyby (CRAF) using X-band transmission.

The sensitivity of this type of search can be improved with modest effort on several fronts. The Deep Space Net has a stated goal of improving frequency standards, transponders, and transmission technology to make use of frequency stabilities of 10^{-17} in periods of thousands of seconds. Should this be implemented, the other noise sources will dominate. The noise from phase fluctuations due to the interplanetary plasma could be further reduced by simultaneous two-band transmission in both uplinks and downlinks. It is estimated that dual-frequency S- and X-band in both links could reduce the plasma noise by at least a factor of 10 in the antisolar direction. Larger improvements would be made if K-band is used. The tropospheric contribution to the phase noise could be reduced substantially by measurement of the column density of water vapor along the receiving antenna beam.

Given the very few opportunities to carry out interplanetary missions in the U.S. space program, it seems prudent to make as many of these improvements as possible and to take every opportunity to continue the search for long-period waves by this technique. The incremental costs on a mission to carry out this research appear small.

6
Programs After 1995

BASELINE PROGRAM

LAGOS: Laser Gravitational Wave Observations in Space

Important opportunities for valuable new types of scientific observations in space exist in the area of gravitational wave astronomy. Ground-based programs for observing gravitational radiation are being pursued actively in a number of countries, as described in Chapter 3. Given proper support, the prospects appear good for detecting signals that give information about several kinds of sources within the next 5 or 6 years. However, the frequency range that can be covered relatively easily in ground-based observations is only from roughly 100 Hz to 10 kHz. With major efforts at isolating the antennas from ground noise and avoiding time-varying gravity gradient effects, high sensitivity can be achieved down to 10 Hz or possibly somewhat below. For frequencies of 3 Hz and below, antennas in space seem essential to achieve sufficiently high sensitivity.

The types of low-frequency gravitational wave signals that are likely to be observable fall into several classes. The only one that is certain to be observed is radiation from various kinds of binary star systems. Sources of this kind include normal main sequence binary stars, contact binaries, cataclysmic variables, neutron star binaries, and close white dwarf binaries. The radiation level expected for the last type is not well known because of uncertainties in the initial distribution of masses and separations. For each type of binary the number of sources in our galaxy is very large, so that even several years of observations would leave many sources contributing in each frequency resolution bin with similar signal strengths. Only a relatively small number of unusually nearby sources would give signals higher enough than the continuum level to be individually distinguishable. Thus the observations would be limited by confusion, since types of antennas suggested so far have rather poor angular resolution.

A very important class of signals that could be present, but about which the theoretical predictions are highly uncertain, is pulses of gravitational radiation resulting from the formation or collisions of very massive (10^2 to $10^5 M_\odot$) or supermassive ($>10^5 M_\odot$) black holes. If stars with masses greater than a few hundred $M_\odot$ form at any time, they are expected to evolve quite rapidly and may well collapse to form black holes. The efficiency of gravitational radiation emission during collapse for some levels of initial rotation has been estimated to be roughly 0.1 percent. However, whether stars with such high masses ever form is a major question. Although a Population III of massive stars is often suggested to have formed early in the history of the galaxies and to have produced the initial heavy element abundance, there is no evidence for current formation of stars more massive than perhaps $150 M_\odot$, even in regions of rapid star formation. Whether conditions were favorable for the formation of much more massive stars in early galactic times is not known.

Another possible scenario for the formation of very massive or supermassive black holes involves pregalactic density inhomogeneities. The probable scale for such inhomogeneities is roughly 10^5 to $10^6 M_\odot$. However, it is not clear whether objects in this mass range that initially were very dilute would fragment, form normal stars, or collapse. Limits on the density of black holes with masses of 10^2 to $10^7 M_\odot$ in galaxies currently are about 1 percent of the closure density. The absence of many such objects in the cores of normal galaxies is not an argument against their existence, since the time for them to spiral into the center due to viscous drag is very long.

The other frequently mentioned possible source for the emission of strong gravitational wave pulses with frequencies below 1 Hz is collisions between very massive or supermassive black holes. If two such objects approach each other with substantial angular momentum, calculations indicate that the efficiency for gravitational wave emission is likely to be fairly high. Whether such collisions are likely to occur even as frequently as a few times per galaxy lifetime is not clear. However, the chances for such collisions might be substantial if rotating supermassive objects existed in pregalactic or early galactic times. Then bifurcation could occur because of the bar instability, followed by collapse of each mass, and a spiraling into a final collision. If as much as 10^{-4} of the mass of the galaxies was put into very massive or supermassive black hole binaries with relatively close separations, the chance of detecting them by their gravitational radiation seems good. Such evidence might include fairly strong periodic signals with a drift toward higher frequencies as the black hole binaries spiral in toward each other.

As is emphasized above, the probability of being able to observe pulses of gravitational radiation due to the formation or interaction of very massive or supermassive black holes is not known at present. If observable, the shapes of such pulses might well be able to give new information about gravitational interactions at distances of the order of the Schwarzschild radius, where general relativity has never been tested before. The absence of such pulses would make it seem unlikely that large numbers of high-mass black holes played a substantial role in the pregalactic or early galactic history of the universe. More exotic sources of low-frequency gravitational radiation such as phase transitions in the early universe also have been suggested. Despite our current limited knowledge concerning sources of low-frequency gravitational waves, it appears important scientifically to carry out a reconnaissance mission in space at an early date to search for gravitational waves over as broad a frequency range as possible in the region below 1 Hz.

The basic approach for laser gravitational wave observations in space is the measurement of changes in the distances between three separate spacecraft with extremely high accuracy. For the baseline case that has been studied over the last few years, the three spacecraft are placed in similar one-year-period nearly circular orbits around the Sun with separations of about 10^6 km. The central spacecraft is in the plane of the ecliptic and has very

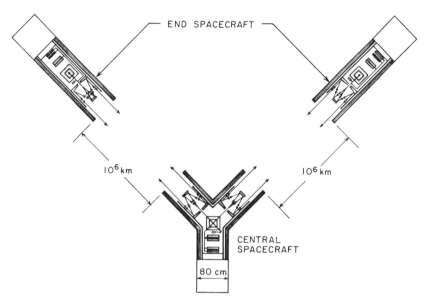

FIGURE 6.1 Laser heterodyne gravitational wave antenna. SOURCE: Courtesy of University of Colorado.

small eccentricity. The other two "remote" spacecraft have almost equal eccentricities of $e = 0.003$ and inclinations of $\sqrt{3}\ e$, but different longitudes for their nodes and perihelia. By choosing the differences in node and perihelion correctly, the two remote spacecraft always will be very close to 90 degrees apart, as seen from the central spacecraft. For these orbits, the distances from the central spacecraft to the two remote spacecraft can be kept equal to about one part in 10^3.

With this geometry, a beam-splitter and two end mirrors can be mounted in free-floating test masses inside the three spacecraft to form a Michelson interferometer. As shown in Figure 6.1, a continuous wave (cw) laser beam is sent to the beam-splitter in the central spacecraft, and then the two resulting beams are transmitted through a pair of 50-cm-diameter telescopes to the two remote spacecraft. There the beams are received by similar telescopes and sent to the end mirrors. The return beams are generated by similar lasers in the two remote spacecraft, which are phase-locked to the received beams. At the central spacecraft, the phases of the beats between the return beams and the central laser are measured as a function of time.

The antenna is sensitive to gravitational waves, since one arm of the interferometer will increase in length and the other will decrease for favorable directions of propagation and polarization of the waves. For 100 mW of transmitted laser power, the expected shot noise limit for the detectable gravitational wave amplitude is $10^{-20}/\sqrt{Hz}$. It appears likely that this sensitivity can be achieved for gravitational wave frequencies ranging from 0.1 Hz to about 10^{-4} Hz. For a 10^4-s integration time, the gravitational wave sensitivity would be 10^{-22}. Useful but lower sensitivity would be achievable over the rest of the frequency range from 3 Hz to 10^{-6} Hz. Thus, a 10^6-km LAGOS antenna would have extremely high sensitivity for detecting and observing astronomical sources of gravitational waves with periods of roughly 0.3 s to 10 days.

The sensitivity of the antenna for short pulses of gravitational radiation is shown in Figure 6.2, along with the expected noise level due to random variations in the power from binaries in our galaxy. Separate curves for the noise level are shown with the close white dwarf binaries (CWDBs) included or excluded, since the amplitude of their contribution is quite uncertain. Also shown are the signal strengths expected for short gravitational wave pulses if very massive or supermassive stars collapse to form black holes at red shifts of $z = 2.5$ or $z = 30$. It is assumed that the efficiency is 0.1 percent, the Hubble constant is 55 km/s/Mpc, and the expansion parameter $q(0) = 1/2$. The interferometer sensitivity will be degraded from the curve shown by a factor $\ln(T/r)$, where T is the time between pulses near that frequency, and r is the pulse length. If frequent pulses are generated at $z = 30$, it is clear from the figure that they would be observable for initial masses greater than about $10^4 M_\odot$, provided that the efficiency is 0.1 percent and the noise from CWDBs is somewhat lower than estimated. For $z = 2.5$, the initial masses would have to be greater than about $3 \times 10^4 M_\odot$.

Laser stability requirements for LAGOS are not as severe as might be expected at first. Although each arm of the interferometer changes its length at a rate of up to about 20 cm/s due to the solar attraction and planetary perturbations, there are no solar system sources of disturbing gravitational forces with periods in the range of 1 s to 10 days that would not be known accurately from their effects at longer periods. Thus, apparent changes in

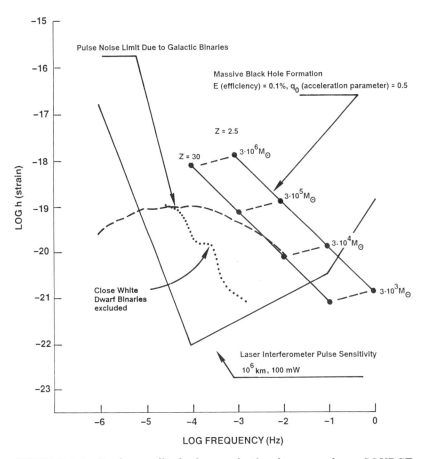

FIGURE 6.2 Strain amplitude for gravitational wave pulses. SOURCE: Courtesy of University of Colorado.

the length of one arm in the period range of interest can be used to correct for changes in the laser frequency. The difference in length of the two arms, after this correction, is much less sensitive to laser frequency variations. Preliminary studies indicate that a laser stability of 1×10^{-13} from 1- to 10-s period, and less stability below and above this range, would be sufficient to achieve the full gravitational wave sensitivity. It appears that the desired stability can be achieved with laser-diode-pumped Nd:YAG lasers, which are being investigated for other space applications requiring high efficiency, long lifetime, and high reliability.

The possibility of locating the LAGOS antenna in geosynchronous orbit instead of in solar orbit also should be considered. For 90 degree separations between the three spacecraft, the interferometer arm length is reduced to 60,000 km. The sensitivity of the antenna would be improved at frequencies above 0.1 Hz and made worse at frequencies below 3×10^{-4} Hz because of the shorter arm length. However, changes in thermal gradients inside the spacecraft probably would generate large spurious signals at a number of harmonics of 1 cycle/day. The effect of orbit perturbations due to harmonics in the Earth's gravity field would have to be considered, and time variations in the mass distribution of the Earth might cause appreciable perturbations. Whether there would be substantial savings in propulsion requirements by placing the antenna in geosynchronous orbit has yet to be investigated.

A third possibility is to reduce the spacecraft separation to about 300 km, and to reflect the light from the remote spacecraft directly back to the central spacecraft instead of using phase-locked lasers. For this separation, many bounces of the light back and forth in each arm could be used with the Fabry-Perot type of interferometer. This approach would then be analogous to that being used in ground-based laser gravitational wave experiments, except for having a much lower frequency range. The main disadvantage is that any perturbations of the test masses on which the mirrors are mounted will cause proportionately larger fractional changes in the interferometer arm lengths than for the larger spacecraft separations. This is expected to be a problem mainly for frequencies below about 2×10^{-3} Hz. However, much improved sensitivity can be achieved from higher frequencies. For 100 bounces in each arm, about 100 times higher sensitivity can be expected if the short noise limit can be achieved. Such a multipass antenna might be desirable if it is decided to focus attention on frequencies above roughly 2×10^{-3} Hz. This antenna could be located either in solar orbit or in geosynchronous orbit.

POINTS: Small Astrometric Interferometer in Space

The proposed instrument is designed to measure relative angular positions of stars to microarcsecond accuracy and would be able to determine the bending of starlight by the Sun to the second order in the gravitational field strength. The first-order deflection is 1.7 arcsec at the solar limb and varies inversely with the distance from the limb. The second-order term is expected to be 11 microarcsec, falling inversely as the square of the distance. The astrometric precision would find application in many branches of astronomy, particularly in a search for planetary systems by measuring the motions of the star about the center of mass of the planetary system.

The instrument concept is a pair of Michelson interferometers of 2-m baseline employing 25-cm-diameter telescopes (see Figure 6.3a). The two interferometers are mounted so their optic axes are approximately 90 degrees apart on the sky (see Figure 6.3b). The nominal 90 degree separation angle is chosen to maximize the number of stars that can be referenced to a given star in establishing an astrometric grid on the sky. The relative position measurement is carried out by executing small changes in the angle between the optic axes. The instrument should be capable of determining the relative position of a tenth magnitude star to an accuracy of 5 microarcsec in an integration time of 10 min. The technical challenge of the instrument is to maintain dimensional stability, especially the angle between the optic axes, against the inevitable thermal drifts and changing gravity gradients over the integration time. Absolute stability is not required, nor possible. The consistent solution of multiply measured angular differences in the astrometric grid can be used to correct for long-term drifts in the instrument. An internal metrology system using frequency-stabilized lasers and fiducial surfaces would be used to determine the optical delay in the interferometers and the angle between the optic axes.

In its present conceptual design the interferometer would take up about one-third of a Shuttle bay. Useful scientific information could be gained by observing from the Shuttle; however, the longer exposure and more stable pointing provided on a platform or free flyer could be used to advantage by this instrument.

Mercury Relativity Satellite

Major improvements in solar system tests of general relativity would be made possible by accurate tracking of a small relativity satellite in a nearly circular orbit around Mercury. One of the primary benefits would be an improvement by 2 to 3 orders of

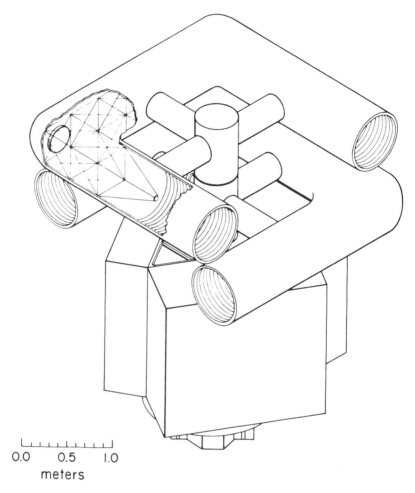

FIGURE 6.3a POINTS: Optical stellar interferometer. SOURCE: Courtesy of Robert D. Reasenberg.

magnitude in our knowledge of whether the gravitational interaction constant G is changing with time. Mach's principle suggests that local inertial properties can be influenced by distant matter in the universe. If so, it could be expected that the effective value of G would decrease as the universe expands. This would cause a quadratic variation with time in the angular position of each planet in its orbit.

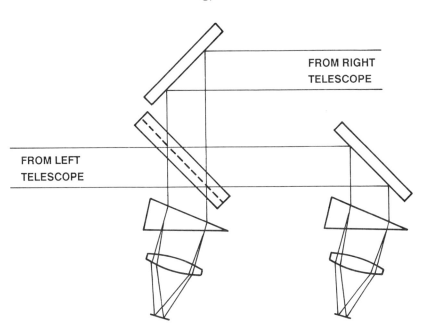

FIGURE 6.3b POINTS: Optical design utilizing double-plate beamsplitter. SOURCE: Courtesy of Robert D. Reasenberg.

Tracking of a relativity satellite also would permit improvements in tests of two other relativistic effects. One is the precession of perihelion for Mercury. This effect is an important test of the extent to which gravitational interaction energy is itself a source of curvature in space-time, as measured by the PPN parameter β. The other test would be an improved measurement of the relativistic time delay for electromagnetic signals traveling along paths near the Sun. The extra time delay depends on the PPN parameter γ, which also determines the deflection of star light or radio waves by the Sun. In addition to these effects, the tracking data would permit an independent measurement of the solar quadrupole moment J_2, which depends on the internal rotation rate of the Sun. And finally, the data would permit a very accurate determination of the gravity field of Mercury up to roughly degree and order 10. This is of great interest to planetary scientists.

For relativity tests, it is necessary to know the spacecraft orbit well enough that the distance from the Earth to the center of mass of Mercury can be determined accurately. For a satellite in a circular polar orbit at 2500-km altitude, dual-frequency Doppler tracking with 5×10^{-15} accuracy can determine the gravity field well enough that the Earth-satellite distance can be converted reliably to the desired Earth-Mercury distance. Phase modulation of the signals at frequencies of up to 50 MHz also is needed in order to measure the range to the satellite with 3-cm accuracy. The resulting overall Earth-Mercury distance uncertainty would be approximately 10 cm. Tracking data are needed for about 3 periods per week of 8 h each over a mission lifetime of 2 to 8 years. The expected mass requirement for a suitable relativity satellite using low-power X-band and K-band tracking is 50 kg.

Recently, a new class of ballistic trajectories has been discovered that would permit sending substantial scientific payloads to Mercury. The trajectories involve swingbys of both Venus and Mercury in order to reduce the final approach velocity. A particularly favorable launch opportunity exists in 1994. The long flight time involved would delay the arrival at Mercury until the year 1998. However, the payload would be large enough that it appears feasible to include a number of separate spacecraft, to be placed in separate orbits. These could include the following: a main spacecraft in a low circular orbit for planetary surface studies; the small relativity satellite; a particles and fields satellite in an elliptical orbit; and possibly one or two small landers. The task group recommends the inclusion of a small relativity satellite on such a mission, or on any other mission that could put the satellite in a suitable orbit around Mercury.

It also is possible to test relativity by using tracking signals from a lander on the surface of Mercury, provided that accurate two-frequency tracking data and long active mission duration can be obtained. This would have the advantage of eliminating the need for determining the gravity field of Mercury. However, the effects of systematic data loss because of the 59-day rotation period have not yet been investigated. Tradeoffs between the lander approach and a small relativity satellite should be considered if it is possible to include a lander on a Mercury mission.

For a 2-year nominal mission lifetime with the small relativity satellite, the expected accuracy for testing the possible change in the gravitational constant is 1×10^{-13} per year. The accuracy for the precession of perihelion and for J_2 are 1×10^{-4} and 3×10^{-8}, respectively. For an 8-year extended mission lifetime, the corresponding accuracies would be 1×10^{-14}, 3×10^{-5}, and 1

× 10^{-8} per year. The expected accuracy for γ from the time delay would be 3×10^{-5}, with only a weak dependence on the mission duration. The present uncertainties are approximately 1×10^{-11} per year for the rate of change of G, 5×10^{-3} for the precession rate, 4×10^{-8} for J_2, and 2×10^{-3} for γ. However, the small quoted present uncertainty for J_2 is from splittings of the roughly 5-min period normal modes of oscillation of the Sun, and an independent direct determination of J_2 is desirable. At the level of 1×10^{-14} per year for possible changes in G, the effect of the comparable rate of mass loss for the Sun by the solar wind would need to be considered.

STARPROBE: Second-Order Gravitational Red Shift Experiment

The experiment proposed is to incorporate a hydrogen maser clock on the mission being planned to explore the plasma around the Sun to make the first measurement of the gravitational red shift to second order in the gravitational field strength (see Figure 6.4). The experiment would provide a direct measure of the PPN parameter β without assumptions concerning values for the other PPN parameters and would have a weak dependence on a solar quadrupole moment. At its closest approach to the Sun the clock will be at 4 solar radii. At this position the first-order gravitational red shift is 5×10^{-7}, while the second-order term is 3×10^{-13}.

The hydrogen maser flown on Gravity Probe A in 1976, which performed a gravitational red shift experiment in the field of the Earth to one part in 10^4, had a long-term stability of 2×10^{-15} in an averaging period of about 3 h. The clocks have been improved since then, exhibiting a stability better than 1×10^{-15} in an averaging time of one day. A hydrogen maser appropriate to the mission has an estimated mass of 35 kg and would consume less than 30 W. The intrinsic stability of the clock is better than needed to perform a 10 percent measurement of the second-order red shift in the time during closest approach to the Sun, about 10 hours. The requirements on spacecraft tracking are not severe; the range error at closest approach should be less than 100 m. However, the effects of the large spatial and temporal variations in the plasma around the Sun must be considered both in the ranging and on the measurements themselves. The angular position of the spacecraft

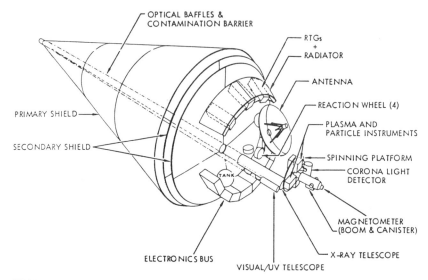

FIGURE 6.4 STARPROBE: Hydrogen maser to the Sun. SOURCE: Courtesy of NASA.

could be measured by very long baseline interferometry (VLBI) using the tracking station antennae.

A possible scheme for carrying out the measurement is to use the first-order Doppler cancellation technique employed in the Gravity Probe A experiment with the additional redundancy of a 4-link system. The 4 links consist of signals originated on Earth and transponded back to the Earth by the spacecraft, and an independent pair originated by the spacecraft and transponded back to the spacecraft from the Earth. The individual carrier frequencies of the signals are chosen to allow a solution for both the first-order Doppler and the first-order plasma terms. Dual-frequency ranging at both K- and X-band will probably be required.

Thermal control of the spacecraft is a key problem in the entire mission; the clock imposes no new special requirements.

High-Precision Principle of Equivalence Experiment

The principle of equivalence experiment can be performed to a level of one part in 10^{18} in a surround that minimizes the gravity gradient problem and produces a "zero-g" environment to the order of 10^{-12} g. One possible satellite in which to perform this experiment would be the refurbished Gravity Probe B (GPB) satellite after 1995. At present, it is planned to recover GPB, and it would require minimum refurbishment for the experiment. GPB is optimized so that the center of mass stays constant as the liquid helium boils away. It contains a complete guidance and "zero-g" system. An experiment performed aboard GPB to an accuracy of one part in 10^{18} would be an improvement of 10^6 over the best existing experiments.

X-ray Large Array/Fast-timing Experiments and Correlation with Gravitational Radiation

The program to develop a large array of proportional counters begun on the Shuttle before 1995 will come to fruition on the Space Station. The scientific importance of fast x-ray timing experiments and their relation to gravitational physics has been discussed in Chapter 5. The full-scale concept is described here.

An array of 512 proportional counters, each 2000 cm^2, for a total effective area of greater than 100 m^2, would be attached to the Space Station (see Figure 6.5). Mechanical collimators would provide a 1-degree-square field of view. The x-ray large array could utilize the Space Station for assembly, operation, and maintenance and would need to handle a data base volume of up to 10^{14} bits, and data rates of up to 10^8 bps. The planned range of energy sensitivity is from 0.25 keV to 50 keV.

ENHANCED PROGRAM

Laser Gravitational Wave Observatory

After an initial reconnaissance mission to observe the general nature of astronomical sources of gravitational radiation, an observatory in space capable of much more detailed studies will probably be needed. Whether such a laser gravitational wave observatory should be planned during the period 1995 to 2015 clearly will depend on the results obtained from the reconnaissance mission. The task group will discuss here the type of observatory that should be considered if substantial evidence is found for pulsed or periodic signals associated with black holes or for other types of sources of major scientific importance.

The main experimental requirements for the observatory are likely to be in two areas. One is that there be multiple antennas for

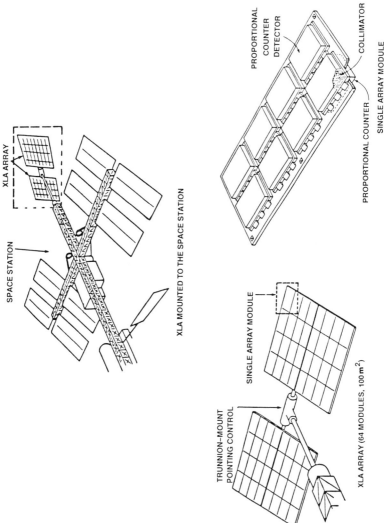

FIGURE 6.5 Large array/fast-timing x-ray detector. SOURCE: Courtesy of High-Energy Physics Laboratory, Stanford University, 1986.

determining the location of pulsed sources, for polarization studies, and for coincidence observations. Quite accurate information on the locations of periodic sources can be obtained with a single antenna by observing when the sources pass through the sharp nulls in the antenna pattern. However, this cannot be done for pulses, unless they arrive frequently and from only a few sources. It generally will be necessary to have observations of the different amplitudes, shapes, and arrival times for signal pulses at different antennas in order to determine source location and polarizations. Whether or not there is any large-scale structure in the source locations is important to investigate. Coincidence measurements with multiple antennas also can substantially increase the sensitivity and reliability of observations made of pulses of gravitational radiation. Three antennas would be sufficient to obtain a large amount of additional information. Separating them by a distance of many gravitational wavelengths is quite feasible if the antennas are placed in solar orbit. One possibility would be to use separations of about 10^8 km in order to achieve the highest angular resolution possible.

The other foreseeable requirement would be to maximize the sensitivity in the frequency region that appears most important based on the observations from the reconnaissance mission. For example, if pulses are seen in the frequency region above 10^{-3} Hz but not at lower frequencies, then the use of roughly 300-km multipass antennas of the Fabry-Perot type probably would be desirable. Also, the achievement of additional power output and improved performance by efficient lasers with long lifetimes for space applications is highly likely by the year 2000. With 1 W of transmitted power, 100 passes per arm, and an integration time of 10^6 s, a sensitivity as high as 3×10^{-26} would be possible for periodic signals in the frequency range of roughly 10^{-2} Hz to 3 Hz. The corresponding pulse sensitivity would be 3×10^{-24} to 5×10^{-23} over the above frequency range. Such high sensitivity would greatly enhance the opportunities for detailed studies of gravitational wave pulse shapes, as well as for discoveries of completely unexpected phenomena.

Reflight of GPB

With improved superconducting quantum interference devices

(SQUIDS) now available and a laser modulation technique developed by Cabrera to modulate the superconducting pickup loop around the gyroscope, it appears possible to reduce the gyro readout noise to a level of 10^{-4} sec of arc per year. With such a system a reflight of the gyroscope could make absolute measurements of several star positions in our galaxy at the two ends of the solar orbit to an accuracy of 10^{-4} sec of arc. This would provide astrometry data on the distance and proper motion of these stars based on absolute gyroscope measurements rather than on comparison of several stars. To make such measurements a 360-degree gyro readout could be developed to an accuracy of 10^{-4} sec of arc. This would have important applications to other astronomical instruments in space.

It is possible that the Gravity Probe B data will be sufficiently interesting to make a second flight, possibly in a nonpolar orbit, very worthwhile. Thus, the capability to bring back Gravity Probe B for refurbishment or to refill with helium in space seems desirable.

7
Technology Requirements

IMPROVED DISTURBANCE COMPENSATION SYSTEMS (DISCOS)

Major improvements are needed in the DISCOS in order to achieve enhanced performance in the laser gravitational radiation observatory. These would include both reduction in the disturbance level below the $10^{-10}/T^2 g/\sqrt{Hz}$-spectral amplitude expected for the initial laser gravitational radiation antenna in space and extension of this performance to periods longer than 10^4 s.

MODERATE-POWER FREQUENCY-STABILIZED LASERS

Frequency-stabilized single radial and longitudinal mode lasers will be useful in many areas of fundamental physics and in astronomy carried out in space. The gravitational wave observations proposed by this study as well as the optical interferometry being considered by the Task Group on Astronomy and Astrophysics will require moderate-power 100- to 1000-mW lasers that will function trouble-free for periods of years. For space applications, it is important that the lasers be efficient; the ratio of the optical output power to the pump power should be larger than in conventional laboratory systems.

A good candidate laser system is the Nd:YAG solid-state laser pumped by laser diodes. Solid-state laser systems using neodymium (Nd) in other host lattices may prove to be even better candidates. The development of high-performance moderate-power lasers is not a major military concern (to the task group's knowledge) and would benefit from funds provided by the space program.

CRYOGENIC CAPABILITY TO TRANSFER HELIUM IN ORBIT

Low-temperature experiments such as the Infrared Astronomy Satellite (IRAS), Gravity Probe B (GPB), the Cosmic Background Explorer (COBE), and the Shuttle Infrared Telescope Facility (SIRTF) are limited by the lifetime of the liquid helium in the spacecrafts' dewar. A very important capability to develop for the future is the ability to transfer liquid helium in space, especially to effect a transfer for low-temperature free flyers. Ultimately, this should be a routine service function of the Space Station, but it could be extremely useful if developed earlier for the Shuttle.

DEVELOPMENT OF CLOCKS

The hydrogen maser now has a long-term stability of better than 10^{-15}. An older version of this clock was flown several years ago on Gravity Probe A to measure the gravitational red shift to an uncertainty of 10^{-4} in a suborbital flight. The hydrogen maser is the logical candidate to use in the second-order red shift experiment proposed for the Starprobe mission. Building a space-worthy improved hydrogen maser is an important technological development program for NASA.

The present development of trapped ion clocks is not as far along as the hydrogen maser but holds the possibility of stabilities of 10^{-17} to 10^{-18}. The continued development of these clocks opens the possibility of a sensitive test of the strong principle of equivalence by intercomparison of clocks with different mixtures of electronic and nuclear energy terms. The intercomparison could first be carried out on Earth, where the solar gravitational potential changes over the course of the year due to the Earth's eccentric motion about the Sun, and later on a spacecraft to sample a larger change in the solar field.

Part B
Science in a Microgravity Environment

1
Introduction

One of the remarkable features of a spacecraft is that it provides the opportunity for carrying out scientific experiments in a microgravity environment. The principal message that the task group wishes to convey in this report is that there are, indeed, strong reasons for exploiting this opportunity.

In the following introductory paragraphs, the task group will describe in fairly general terms the rationale for research in microgravity science, and outline what it feels are some important constraints that must be observed to implement a successful program. The central portion of this report is devoted to illustrative descriptions of a few scientific projects. The report concludes with a summary of the task group's recommendations.

The task group stresses at the outset that a NASA program in microgravity science must be fundamentally different from other space-based projects in its underlying motivation and structure. Unlike, for example, a space telescope or a gravity probe, which provide unique windows to the world beyond the surface of our planet, microgravity science provides just one especially new way of looking at a vast range of phenomena. Thus, general areas of opportunity are emphasized here rather than specific projects. The task group will describe a number of such projects in some detail.

The task group's primary purpose in doing this is to illustrate why a long-term program is interesting and useful.

There are two related reasons for interest in microgravity from the most fundamental scientific point of view. The more mundane is that gravity often induces effects that obscure essential features of the phenomena being studied. The denser product of a phase transition or a chemical reaction settles to the bottom of the sample before all the intrinsically important processes have gone to completion. Or the homogeneity of the sample is destroyed because its own weight compresses it more at the bottom than at the top, thus impairing measurements. Such effects occur commonly in experiments in hydrodynamics, in novel materials, in biological systems, and elsewhere. We are seeing enormous progress in such areas largely because new experimental and computational techniques allow us to obtain and interpret information in the kind of detail undreamed of just a few years ago. The examples described later in this report illustrate a number of cases where elimination of gravitational effects leads to a major simplification.

The second, more dramatic reason for interest in microgravity science is the real possibility of discovering completely new phenomena. Obviously, we cannot point with any certainty to areas where such discoveries might be made, but some fascinating hints do exist. In the examples section of this report the task group will describe the theoretical prediction of a new state of matter called a "fractal aggregate"—an extremely tenuous form of a space-filling solid whose effective dimensionality, in a mathematical sense, is less than three. Such a material would be intrinsically unstable in a gravitational field, but an understanding of its properties in a freely suspended state might be profoundly meaningful. In addition, there are a number of other situations discussed below where removal of gravitational effects could conceivably reveal entirely new and unexpected phenomena.

In presenting specific examples of how a microgravity environment may be used to carry out significant scientific research, the task group will focus on several broad categories of phenomena distinquished by the manner in which gravitational forces affect the states or processes involved. First is a set of situations in which careful measurements of a system's equilibrium properties are desired, but where the system's state of equilibrium in a gravitational field is different from what it would be without that field. The classic example of such a situation is liquid helium near its lambda point or, more generally, a very wide variety of systems that undergo dramatic changes in their thermodynamic properties in the neighborhood of second-order phase transitions. These so-called "critical phenomena" have been of immense interest during recent years because their theoretical interpretation has a bearing on scientific topics ranging from applied metallurgy to elementary particle physics and cosmology. The theory of critical phenomena has become detailed and sophisticated. To test its limits of validity, we now need extremely accurate measurements. A practical limit to the accuracy of earth-based experiments is imposed by the gravitationally induced nonuniformity of the equilibrium state of the system; the pressure is greater at the bottom than at the top. As a result, only a very small part of the system can be held precisely at its critical point. In other words, gravity limits the effective site of the experimental sample by deforming it, and thus limits the precision of measurements. Planning for a lambda-point experiment in space is now well under way and is described below. Also discussed below is a similar situation in which gravitational deformations of granular media inhibit accurate observation of failure mechanisms in such materials.

The second category of examples described here includes situations in which gravity causes stationary states or processes not just to deform but actually to become unstable. The gravitationally induced collapse of a fractal aggregate is an especially interesting example of such a situation. Another example in this category is a phase-separating mixture of fluids. Here the fundamentally interesting diffusive interactions between emerging droplets are obscured by buoyancy-induced flow. In this example we are dealing with a class of phenomena involving *non*equilibrium states of matter, a largely underdeveloped field of scientific enquiry. The interest here is in *processes* rather than simply states of equilibrium, and the potential relevance of the microgravity environment stems from the fact that gravity makes it difficult or impossible to disentangle one process from another. In the case of the fluid mixture, the evolution of a precipitation pattern under conditions of diffusion control is of great fundamental interest; but simple diffusion ordinarily is disrupted by sedimentation—the lighter precipitates float to the top and the heavier ones sink—before quantitatively satisfactory measurements can be completed.

A final category of examples includes strongly nonequilibrium processes—very much at the frontier of modern research—in which the simplification achieved by eliminating extraneous gravitational effects might make it easier to discover the fundamental principles that are involved. The specific examples presented here are surface-tension-driven flows in hydrodynamics, pattern formation in crystal growth, combustion of particulate clouds, and electrically driven flows, all of which figure in current plans for research on the Space Shuttle. However, the task group believes that the potential for microgravity research in this general area is much broader than has yet been appreciated; in particular, many of the basic issues in this area are directly relevant to biological processes. The difference between this category and the previous one is that this category treats systems that are being driven persistently away from equilibrium and that exhibit complex dynamical behavior. Basic scientific questions include how to predict whether this behavior will be regular or chaotic, what spatial patterns will be formed, and whether those spatial patterns themselves will be stable. A common feature of many phenomena where such questions are relevant is that underlying symmetries may be broken by gravitational forces. The front-to-back symmetry of the environment of a propagating flame front or the tip of a growing crystal may be broken by buoyancy; or an interesting instability of a hydrodynamic system may actually be stabilized by gravity because of symmetry breaking. In all of these cases, gravitational effects perturb, obscure, or completely destroy the dynamical behavior of interest.

The focus of this report is entirely on basic science. The task group will not try to evaluate or make recommendations concerning specific technological applications. In general, however, the task group believes that ultimately the greatest commercial benefits of microgravity science will be gained from interactions between basic and applied research, rather than from direct efforts to manufacture or process materials in space. As we gain understanding of the conditions under which gravity fundamentally alters structures and dynamic processes, novel applications are likely to develop. The task group views the creative interplay of basic and applied science in microgravity to be of the highest priority, with the benefits likely going in both directions. For example, semiconductors are the best understood materials in the world today, largely because of the commercial importance of the transistor. In turn, the transistor was invented on the basis of new quantum mechanical understanding of the dynamics of electrons in solids. The task group believes that we should be looking for such breakthroughs in technology as well as science when exploiting the microgravity environment.

Let us turn now to some issues that seem to the task group to have major implications for the implementation of a national program in microgravity science. First is the role of the computer, which has revolutionized the way in which research is performed, and which must be taken into account in deciding whether and how to carry out experiments in space. It seems highly likely that within the next decade—well before a microgravity facility is apt to be in full operation aboard a space station—we will be able to perform accurate numerical simulations of many of the systems discussed in this report. Fully three-dimensional simulations of fairly complex hydrodynamical flows, flame fronts, or solidification patterns seem to be only slightly beyond the reach of current supercomputers and numerical algorithms. They will almost certainly be possible within a few years.

Does this mean that experiments in space will be irrelevant? In some cases, the answer to this question may be yes. If we are quite sure of the physical ingredients of the simulation model, and have been able to test the accuracy of the computer code against ground-based experiments, then there would seem to be little sense in spending the effort and resources necessary to fly an experimental "analog computer" on a spacecraft. On the other hand, if there is fundamental uncertainty about the physical principles, then numerical simulation may actually be the key to making microgravity experiments feasible and meaningful. The ideal microgravity experiment will be one in which the computer has been used, not just to design the apparatus and control its operation, but also to make the observations more meaningful by providing quantitative predictions. Thus, the new level of interaction between theory and experiment made possible by the computer provides a rationale for microgravity science that would not exist otherwise. The computer can help investigators design experiments effectively to highlight critical features of the phenomena being studied; and properly designed experiments can test and refine theoretical hypotheses in a way that, ultimately, will lead to better simulations.

The last issue that the task group will address here concerns what might be called the "infrastructure" of microgravity science. As noted earlier, this field is quite different from other, more uniquely space-related areas of research. The differences pertain both to the nature of the research and to the way in which it is carried out. A thoughtfully implemented microgravity program may eventually have an impact on all fields of science and technology that are concerned with properties of materials. Thus, few areas of scientific activity should be completely uninterested in it. Conversely, few of these areas of activity are completely dependent on a microgravity facility in order to pursue their investigations. In contrast, astronomy, astrophysics, and relativity are coming to rely heavily on space-based observational facilities. All of the research problems discussed here have emerged in fields that have existed outside of space science and will continue to exist with or without a laboratory on a space station. The task group has argued that a microgravity facility may be quite useful in many of these areas and might even be crucial for certain special projects. In addition, it argues that, in order to take advantage of these opportunities, we must recognize the special needs of the scientists who will do the work.

A crucial ingredient of a successful microgravity program, the task group believes, should be a well-equipped, scientifically staffed research center that functions much as a national laboratory does. The purpose of this center would be to provide leadership in both the scientific and the technological aspects of microgravity research. Such a center should support full-time, in-house activities and should also serve scientists from universities, industry, and other national laboratories by enabling them to carry out microgravity research in a timely, cost-effective manner. This proposal is intended to encourage a broad range of the nation's most capable scientists to commit their efforts to microgravity research. Microgravity experiments generally will be relatively small but crucial parts of broader research programs in which related problems are being attacked using different techniques. If they are to work at all, the microgravity projects will have to fit comfortably into this overall scheme, both with regard to timing and with regard to the amount of effort required. Thus, the costs to investigators in both time and resources will be crucial factors in their decisions about whether to attempt experiments in space. A visible and effective national research center would address many of these practical questions. Its permanent staff could relieve principal investigators of the need to master all of the special techniques required for space experiments and, with experience, develop efficient and flexible procedures that will make the microgravity laboratory accessible to those who are best able to take advantage of it for basic scientific purposes. This staff would also be able to help investigators shorten the path to success, perhaps by using simple, quickly implemented experiments as opposed to major long-term projects.

In summary, the task group believes that a microgravity research facility can justifiably be made a significant component of plans for the future of space science. Although the microgravity program may ultimately have a major impact on technology, its principal rationale should be its importance for basic research. Successful implementation of such a program requires extensive ground-based preparation of individual projects, starting with the careful definition of the basic questions to be addressed, and continuing with detailed simulations and design studies for experiments. This is a serious effort for which, the task group believes, the rewards will be substantial.

Finally, the task group views the active participation of outstanding young scientists as a crucial element in the long-term productivity of the microgravity program.

2
Gravity-Sensitive Systems at Equilibrium

CRITICAL PHENOMENA

Near their critical points, "many-body" systems exhibit long-range order that greatly exceeds the range of the interparticle interactions. The characteristic length over which the long-range order persists is called the correlation length. If all of the thermodynamic variables of a many-body system can be held sufficiently close to those defining the critical point, the correlation length can approach one millimeter. This long correlation length produces anomalies in the thermodynamic, transport, and structural properties of the system. These anomalies are remarkably independent of the choice of system. Experimental and theoretical studies of these anomalies offer an exciting approach to the general "many-body problem," which has challenged physics for a very long time.

The same long-range order exhibited in many-body systems near their critical points also exists in systems that have "second-order" or cooperative phase transitions. A transition especially suitable for experimental study is the so-called lambda (λ) transition between normal and superfluid liquid helium, as discussed later in this section.

Recent interest in critical phenomena has been stimulated by the new theoretical understanding of these phenomena achieved by

Kenneth Wilson, who applied the Renormalization Group Technique to the problem. This work led to his being awarded the Nobel prize. Even though Wilson's work produced specific values of the coefficients and exponents that characterize the critical anomalies, the experiments have not yet been able to test the finer details of the theory. Accurate experimental values of the predicted quantities require extremely precise measurements under tightly controlled conditions (temperature, pressure, density, and so on). Otherwise, the experiments are unable to distinguish between conflicting theories. In some experiments the temperatures of the sample must be uniform, measured, and controlled to one part in 10^{10}. On Earth the necessary uniformity cannot be achieved throughout the sample because of strain produced by the Earth's gravitational field. Therefore the experiment must be done in a reduced-gravity environment to permit definitive measurement.

Projects leading to space experiments on matter near critical points or passing through cooperative phase transitions are now proceeding in several universities and at the National Bureau of Standards. Each of these projects involves collaboration with one of the NASA flight centers. Examples of these experiments follow.

One experiment measures the decay rates of critical fluctuations of a simple fluid (xenon) very near its critical point by observing the light scattering from thin samples of the fluid. The experiment cannot be performed on Earth because the very large compressibility of any simple fluid at its critical point will lead to a large density gradient in the sample due to the sample's own weight. In a low-gravity environment the density gradient will be greatly reduced and it will be possible to keep the entire sample within one microdegree Kelvin of the critical temperature. Under these conditions the experiment will provide a definitive test of the theory of decay rates.

Another experiment measures the viscosity of xenon near its critical point to develop theoretical models of transport properties of critical fluids and to provide another test of Renormalization Group predictions. Still another experiment studies transport properties, i.e., thermal conductivity and shear viscosity near the liquid-vapor critical point of ^{3}He and very near the critical point of ^{3}He/^{4}He mixtures.

Of all the critical-phenomena experiments now in progress, the one nearest to flight is the so-called lambda point experiment, which measures the heat capacity of liquid helium through its lambda transition from its superfluid phase gravitational levels down to the 10^{-5} to 10^{-6} g range. Evolution of some of these experiments into the twenty-first century will call for "nano" gravity, i.e., 10^{-9} g and below. Technology development to meet these and other requirements of the microgravity and nanogravity experiments should be vigorously pursued.

This task group believes that the microgravity environment offers opportunities for important scientific advances in the testing of fundamental theories and the discovery of new phenomena and new states of matter. In particular:

1. Our ability to test fundamental concepts such as Renormalization Group Theory can be greatly enhanced in the absence of gravity-induced nonuniformities.

2. Our understanding of nonlinear nonequilibrium phenomena (fluid flow, condensation, solidification, combustion, and a wide range of related dynamical processes) can be significantly advanced by experiments under conditions where the underlying behavior is not masked by the effects of gravity.

3. New static or dynamic states of matter that do not exist in normal gravity may be created and investigated.

In order to take advantage of these opportunities, we must recognize certain unique features of the microgravity program: its relevance to an extremely broad range of scientific activities and its resulting need for versatility and responsiveness in its operation.

Now that the technology for using liquid helium in microgravity is being developed to the point where we can operate very close to the lambda transition, it will be possible to address a host of other fundamental problems. One such problem is the direct determination of the correlation length for liquid helium. Another is the construction of a weak link for superfluid helium that would allow the study of Josephson effects, which heretofore have been seen only in superconductivity. A weak link is a connection between two helium baths that depresses the wave function of the order parameter but allows the phases on either side to remain related; such a connection will likely have to be of the order of the correlation length in size. Attempts have been made to construct a weak link in ground-based experiments, but without apparent success—perhaps because the correlation length is of the order of angstroms at temperatures that can be attained on Earth. In a microgravity environment the correlation length could approach the size where a weak link could be constructed in a machine shop. It would be fascinating to see what effect the various forms of dissipation near the lambda transition would have on the tunneling rate in such a junction. Indeed, friction could be regulated by changing pressure or adding ^{3}He. Another situation that will change with access to microgravity is the structure of the cores of quantized vortices. The cores will become macroscopic in size close enough to the lambda transition, profoundly altering their dynamics. For example, vortex waves, vortex-vortex interactions, and other phenomena will be altered as the sample reaches the previously forbidden region close to the lambda point. It is entirely possible that a way may be found to introduce tracers that will allow the three-dimensional motions of these vortices to be visualized. We would then have a direct understanding of the true nature of quantum turbulence.

Finally, the task group emphasizes the new technology that will emerge as these critical-phenomena experiments are prepared for space. In order to take full advantage of reduced gravity, spacecraft instrumentation is being built that reaches new levels of sophistication. For example, the group performing the lambda point experiment has already invented and built thermometers that will measure the temperature of the liquid helium sample to one part in 10^{10}.

They have also developed the calorimeter and control system that will maintain the temperature to that precision. This alone is a technological feat. Similar innovations will probably attend the development of other experiments.

MECHANICS OF GRANULAR MEDIA

Consider the material behavior of cohesionless granular materials such as sand, flocs, or powders. These materials are very weak compared to other engineering materials, and therefore have low failure stresses. Postfailure behavior is important in a variety of scientific and engineering applications, including chemical processing of catalytic powders, earthquake engineering, and geophysical catastrophes such as eruptions and landslides. The presence of gravity represents a nearly insurmountable obstacle to the measurement of failure criteria, failure mode, and microscopic failure mechanisms on the one hand, and postfailure behavior on the other. This is so since, in a sample of sufficient size that continuum properties can be measured, the axial stress due to gravity has caused a stress gradient in the sample. This in turn leads to an inhomogeneous material, since some portions are postfailure and some prefailure. Some Spacelab experiments have already been proposed to study these materials. The objectives of the experiment are to make measurements of the stress-deformation behavior of these materials under low confining stresses. Conventional triaxial compression tests with continuous measurement of load and strain are planned. Furthermore, local microscopic photographic measurements of the motion of individual particles and grains will give insight into the grain-level mechanisms of failure. These results will be useful for formulating and testing nonlinear constitutive equations for such materials.

3
Gravitational Destabilization of Stationary States

MECHANICS OF SUSPENSIONS

Study of the effective rheological properties of suspensions in a quantitative sense dates from Einstein's thesis (1905). Because of the industrial and technological importance (slurries, separations, colloidal dispersions, flocculation, blood flow, and so forth), suspension mechanics has been intensively studied for the last 80 years. One quantity of interest is the effective shear viscosity of a suspension of neutrally buoyant particles. Einstein showed that for dilute suspensions of spheres the shear viscosity depends linearly on the volume fraction. Data on such systems deviate from linear behavior, with much scatter at high concentrations; there is still no convincing theoretical description of such behavior. More complicated systems (anisotropic particles, emulsions whose dispersed phase can deform, polymer molecules whose conformation can change dramatically, and colloidal systems influenced by electrostatic and London forces) are even more challenging.

Frankel and Acrivos suggested that the viscosity diverges at a critical concentration ϕ_c as $(1 - \phi/\phi_c)^{-1/3}$, where above this critical concentration the suspension can no longer be sheared without dilation. The "critical exponent" is in reasonable agreement with experiments, but because of the scatter in the data, both the leading constant and the critical concentration vary considerably. Recently, large-scale numerical simulations by Brady and Bossis of sheared, *two-dimensional* layers of spheres at high concentrations show that effective properties do diverge, and that this divergence is a result of the formation of large transient clusters of particles. The divergence is associated with a percolation threshold that causes the correlation length of the clusters to diverge. Thus, as the critical concentration is approached, the finite size of the apparatus comes into play, with the accompanying variation from apparatus to apparatus. In order to probe the region near ϕ_c, it is necessary to use a large apparatus. However, no system is neutrally buoyant; small mismatches in density, small temperature gradients, and so forth, will have large effects as the sample size increases. We can illustrate this with the following simple argument.

A measure of the effect variations in density would have on such a suspension of characteristic dimension is given by the sedimentation velocity $g\Delta\rho d^2/\rho\nu$. Suppose the density variation in $\Delta\rho/\rho$ is controlled to 1 percent and the fluid is a viscous solvent with kinematic viscosity $\nu \simeq 1$ cm^2/s. Then if the sedimentation velocity is to be small—less than 1 cm/s, say—then $d^2 << \rho\nu/\Delta\rho g$ in cgs units.

When $g = 1$, we have $d << 3$ mm, which is far too small a characteristic scale to probe for percolation behavior near ϕ_c. In practice, greater control of density is possible, but working with more viscous solvent presents mechanical problems with the high effective viscosities encountered near ϕ_c.

Obviously, working with smaller gravitational levels allows d to be increased without encountering significant sedimentation problems, thus enabling us to work closer to the critical concentration.

SHEARED SUSPENSIONS OF GRANULAR MATERIALS

Suspension mechanics is the description of the rheological behavior of all sorts of suspensions, but the focus over the past 200 years has been on rigid particles suspended in liquids or gases. The theory of systems where the interstitial fluid is gaseous dates to Coulomb's treatment of the static behavior of granular media in 1776. For solids suspended in liquids, the behavior is heavily influenced by the nature of the liquid; bulk rheological behavior reflects the mechanical properties of the liquid, modified by particulate matter. This is in sharp contrast to the situation with granular materials where the solid is suspended in a gas. In this situation the primary sources of stress involve direct contacts between particles. One mechanism for producing stresses resembles molecular momentum transport in gases, where the transport process involves the exchange of particles between layers in relative motion. This mechanism is less important as the volume fraction of solids increases. A second mechanism involves collisions between particles from layers in relative motion—something akin to molecular transport in liquids. The third mechanism involves forces between particles at points of sustained contact. At large volume fractions, the contact stresses will be dominant for low rates of deformation, and the collisional stresses will dominate at high rates. Although the latter two mechanisms resemble momentum transfer in fluids, there is no analog of thermal agitation. Deformation causes particle movement and the random components of the particle velocities arise from collisions between particles. Consequently, these effects die out rapidly as bulk motion ceases.

According to theories now under development, the balance equations consist of conservation laws for mass, linear momentum, and a "pseudo thermal energy." Various versions of the theory differ in the form of the stress tensor, their treatment of the rate of conversion of the "pseudo thermal energy" to true thermal energy by collisions, and the conductivity for "pseudo thermal energy." Since the particles themselves vary in their density and elastic properties, have rough surfaces, and need not be spherical, there are many interesting aspects to any experimental or theoretical study. It is crucial that the various mechanisms be separated in an experimentally meaningful way, but in the terrestrial environment particle behavior is dominated by gravitational effects, making such separations impossible. The influence of the particle contacts with any boundary is confounded with that due to particle-particle contacts, and vice versa. Rheological studies in a microgravity environment offer clear advantages over terrestrial studies since we could expect to measure and test the tenets of any theory separately.

GROWTH OF CONDENSATES IN SUPERSATURATED SYSTEMS

Common examples of the processes referred to here are the formation of droplets of fog in a supersaturated vapor or the precipitation of impurity-rich particles in metallic alloys. Processes of this kind occur in a wide range of chemical, physical, and biological situations. These processes have never been adequately understood from a fundamental point of view, despite having been known since scientists first looked through microscopes, and despite their importance in determining the structure and properties of many substances of technological interest.

The conceptual difficulties are twofold. First, there is the problem of nucleation: the question of how droplets form after the system has become thermodynamically unstable. This is a challenging problem in nonequilibrium statistical mechanics that continues to be of great theoretical interest. It is in the second stage of the process, in the growth and coarsening of the precipitate, where microgravity may play a role.

The second stage is characterized by the competition between several mechanisms. One of these mechanisms is that molecules of the condensing phase tend to evaporate from the smaller droplets and recondense onto the larger ones, thus increasing the characteristic size of the objects in the precipitation pattern. Another competing mechanism is that droplets may merge with one another. In both mechanisms, the driving force is surface tension; that is, the system evolves toward states of lower surface-to-volume ratio and thus lower surface energy. At early stages, this coarsening process is ordinarily controlled by diffusion of material from one droplet to another. At later stages in fluids, surface tension can drive hydrodynamic motions in which material flows from one region to another—for example, through channels formed by merging droplets. In solid systems, further mechanisms are associated with elastic forces, interactions with defects, grain boundaries, and the like. Even the simplest diffusion mechanism presents a major theoretical challenge, especially when the condensing phase occupies a nonnegligible fraction of the volume of the system and the droplets are close enough to each other so that deformation of their shapes may be important. At present, the state of our understanding of these processes is so rudimentary that a metallurgist, using the most modern tools of analytic microscopy, cannot tell whether he is looking at an initial transient or a nearly equilibrated, late stage of a precipitation pattern. In other words, he has no fundamental scientific understanding of how to classify and control these processes.

One special need in this research area is careful experimentation using the simplest possible physical systems and observation of their behavior over the longest possible times. It is here that gravitational effects enter the picture. The simplest model systems are fluids—binary mixtures that undergo phase separation when subjected to changes of temperature or pressure. Fluids can be made very pure, and automatically have none of the complexity associated with crystalline anisotropies or defect structures. They are, however, subject to gravity-induced flows when they become spatially inhomogeneous on sufficiently large length scales. Even when attempts have been made to use working fluids for which the two separating phases have nearly the same density, it has not been possible to prevent precipitation patterns from breaking down under gravity for long enough times to span the theoretically interesting range of behavior.

It is quite possible that this observational problem could be solved in a microgravity environment; thus, the results would be of fundamental significance. This class of experiments probably would require very low average acceleration over periods of the order of days, but may not be sensitive to small accelerations at higher frequencies. On the other hand, it is not clear whether other symmetry-breaking perturbations, such as wall effects or electrostatic forces, could cause major problems.

FRACTAL AGGREGATES

An exciting area of research is the study of systems that can be adequately investigated only in a low-gravity environment. An example of such systems is fractal aggregates, objects whose mass $M(L)$ varies as some power of their size, viz., L^D. Here D is the fractal dimension that lies between 2 and 3 and that depends on the growth kinetics of the aggregate. Since $D < 3$, these tenuous solids become more diffuse as they grow and represent a new state of matter. Unfortunately, gravitational forces limit the growth of aggregates in solution because of sedimentation, which depends on the size of the experimental volume, the viscosity of the medium in which the aggregates grow, and other factors. While the viscosity, and hence the sedimentation time, can be increased, undesirable physical and chemical effects occur that fundamentally alter the aggregation process. At present, it has been possible to verify the fractal growth law $M \simeq L^D$ only over 2 orders of magnitude in size; significant questions remain unanswered. If experiments could be performed in a microgravity environment, one could investigate the fractal growth law as well as the scaling dependence of correlation functions over a considerably wider range of system parameters. In addition, the complicating influence of hydrodynamic flow on the aggregation process would be eliminated by disposing of gravitational effects.

A second major advantage of microgravity experiments on aggregate growth is the elimination of mechanical collapse of aggregates due to their own weight. Theory predicts that the maximum size of mechanically stable aggregates L_M is of order 10 to 100 μm for clusters of 200-Å gold or silica particles. Thus, even aggregates less than 0.1 mm in size collapse to "zero volume pancakes," since the collapsed density vanishes,

$$M/L^3 \simeq L^{D-3} \to 0 \text{ as } L \to \infty.$$

To avoid self-destruction of large aggregates, work must be done in a nearly weightless environment. Theory predicts that L_M varies inversely with g: $L_M(g) = L_M(1)/g$. Thus, for gravity reduced by 10^5, L_M is of order 10 to 100 cm, a desirable range for experiments.

Once stable aggregates of macroscopic size have been obtained, a host of exciting experiments on a single fractal become possible: elastic, electrical, and thermal transport properties can be measured. At present, these measurements are impossible because the medium dominates bulk measurements on aggregate ensembles.

Another interesting system to study is a fractal gel in which individual aggregates join together by percolation to form an infinite cluster. Such gels are likely to have unique properties. No doubt, other examples of fractal structure occur in nature, possibly in macromolecular assemblies.

4
Systems Far from Equilibrium

SOLIDIFICATION PATTERNS

In dendritic crystal growth, extraneous gravitational effects obscure the underlying physics of a pattern-forming process. The term "dendrite" refers here to tree-like patterns, which are most familiar in the form of snowflakes, and which are of special technological importance because of the role they play in determining metallurgical microstructures.

Conceptually, dendritic solidification of a pure substance is one of the simplest examples of pattern formation in nature. Simplicity, in this case, means that we are fairly sure what the elementary physical model of the situation must be. We consider a solid forming within an undercooled liquid, for example, an ice crystal growing inside a sample of pure water cooled to below its freezing point. Growth under these circumstances is controlled almost entirely by the rate at which the latent heat generated at the solidification front can diffuse away through the liquid. A statement of thermodynamic boundary conditions at the liquid-solid interface completes the specification of the model. We have good reason to believe that, were we able to solve the mathematical equations implied by the above sentences, we would predict the generation of a rich variety of snowflake-like shapes quite similar to those observed in nature.

In fact, this solidification problem turns out to be anything but simple. Along with certain roughly analogous problems in hydrodynamics (Rayleigh-Benard convection; viscous fingering) and in chemistry (reaction-diffusion problems), it seems to epitomize the fundamental question of how complex structures are generated when initially structureless systems are driven strongly away from equilibrium. Even in the superficially simplest of these models, we are only beginning to be able to predict some special properties of emerging patterns. In general, we are still unable even to determine whether an emerging structure will be regular or intrinsically chaotic. Here, as in any developing scientific field of investigation, what is needed is an interplay between theoretical work and carefully controlled experimentation. The most useful experiments are always the simplest, that is, the ones in which the phenomena of fundamental interest are least obscured by extraneous effects. The problem of dendritic solidification is an especially clear example of a situation where gravity is the leading culprit in producing extraneous effects.

A basic feature of dendritic solidification as it occurs under the conditions described above is that at the speed, v, at which the leading tip of the tree-like structure grows, the radius of curvature, ρ, of this tip, and the spacing, λ, of the first side branches that appear behind it are all well-defined, reproducible functions of only the undercooling of the fluid. Thus, the classic experiments of Glicksman and his colleagues have been aimed at measuring v, ρ, and λ. It is important for a variety of reasons that these observations be made at small undercoolings, that is, at low thermodynamic driving force. In this regime, v is small and there is little chance that poorly understood departures from thermodynamic equilibrium at the interface can play much of a role. Also, in this regime of slow growth ρ and λ turn out to be large, and thus one can achieve considerable accuracy in measuring these geometric quantities by photographic means.

At this point, the gravitational effects enter. The latent heat released by the tip of the dendrite warms the fluid and causes it to rise. If the dendrite is growing upward, this gravitationally induced convection opposes diffusion of heat away from the tip, and growth is retarded. Conversely, downward-growing dendrites are accelerated. Unfortunately, the effect is most serious at small undercoolings where length scales are large and viscous forces are relatively ineffective. As a result, accurate velocity measurements can be made only when undercooling is high enough. However, sharp pictures can be taken only at low undercooling. A fully satisfactory comparison of experiment with theory requires simultaneous measurements of velocity and tip radius, but there is only an inconveniently small range of intermediate undercoolings in which this can be done. Obviously, some method for appreciable reduction of this gravitational effect would be most useful. A microgravity experiment for the Space Shuttle is now in the planning stage.

SURFACE TENSION AND CONVECTION EFFECTS

Surface-tension-driven phenomena arise whenever a free surface interacts with a field (temperature, magnetic, electric fields) that influences surface tension.

A fluid heated from below involves a stability problem, with a threshold temperature difference ΔT required for convection. A fluid heated at the sidewalls, on the other hand, yields convection no matter how small ΔT. Defining α as the thermal expansion coefficient, ν the kinematic viscosity, κ the thermal diffusivity, ρ the density, σ the surface tension, d and w the depth and width of a fluid layer, and μ the viscosity, Davis and Homsy have shown that in the former case the problem is governed by no less than seven dimensionless groups. These are: the Rayleigh number $\text{Ra} = g\alpha\Delta T d^3/\nu\kappa$, the Marangoni number $\text{Ma} = (-\partial\sigma/\partial T)\Delta T d/\rho\nu\kappa$, the Capillary number $\text{Ca} = (\mu\kappa/\sigma d)$, the contact angle γ, the Prandtl number $\text{P} = \nu/\kappa$, the Bond number $\text{B} = \Delta\rho g d^2/\sigma$, the aspect ratio $\text{A} = d/w$, and thermal boundary conditions at all surfaces. For other fields such as electromagnetism, similiar groups arise. Gravity enters the problem through Ra (buoyancy-driven effects within the fluid) and B (density differences, $\Delta\rho$ at the interface). Ca is a measure of interfacial deflection due to convection. Different aspects of the phenomena associated with free surfaces involve gravitational acceleration in varied ways.

Minimizing Buoyancy-driven Effects

In order to do an experiment that is *essentially* dominated by surface-tension-driven convection far into the nonlinear regime at high Marangoni numbers, we must have Ma >> Ra, or $d^2 \ll (-\partial\sigma/\partial T)/(\partial\rho/\partial T)g$. Using typical values of $-(\partial\sigma/\partial T) = 7 \times 10^2$ dynes/cm K, $-(\partial\rho/\partial T) = 1 \times 10^{-3}$ g/cm^3 K, $g \simeq 10^3$ cm/s^2, $d \ll 3$ mm. It is well known that for most fluids, we must go to submillimeter length scales before minimizing (but not completely eliminating) buoyancy effects. Such experiments in thin layers have been conducted by several groups, but with some degree of difficulty. Furthermore, most measurements at such length scales are limited to global properties (average surface temperature, average heat flux, etc.). Field information (e.g., velocimetry or thermal profiling) cannot be obtained when working at such small length scales. A spacecraft experiment is currently under development in an attempt to measure such field information in domains whose characteristic length is of the order of centimeters rather than millimeters. Microgravity studies of surface-tension-driven flows thus present some opportunities, but also present some complications regarding the effect of g-jitter on interfacial stability and integrity.

Minimizing Free Surface Deflections

Free surface deflections as a result of fluid motion make the convection problem a free-boundary one. Challenging numerical difficulties arise if such free-boundary problems are treated in generality. However, for most fluids, Ca = $(\mu\kappa/d\sigma) \simeq 10^{-4}/d$, where d is in cgs units. Thus Ca is small for any reasonable length, d, and steady flows may be computed for comparison with experiment by domain perturbation. However, instability modes may be intimately linked with free-surface deformability and three-dimensional flows. As a result, experiments involving finite-amplitude traveling waves are the only way of probing nonlinear instabilities; computations of such three-dimensional, time-dependent, free-boundary problems at high Marangoni numbers are simply too difficult.

MINIMIZING THE EFFECT OF GRAVITY ON FREE SURFACE SHAPES

In order to minimize the effect of gravity, the Bond number, B, must be much less than unity. For most fluids in 1 g,

$$B \simeq (1)(10^3)(d^2)/50 \simeq 20d^2$$

Again B << 1 implies d^2 << 0.05 cm^2, or d << 2 mm, i.e., length scales of the order of millimeters. Here microgravity experiments offer one means of suppressing static deformation of free surface shapes.

ELECTROKINETICS

The subject known as electrokinetics lies at an intersection between fluid mechanics and colloid chemistry since charged interfaces with diffuse layers of space charge produce phenomena different from those encountered in traditional fluid mechanics. Motion can arise in response to an imposed electric field, as in electrophoresis and electroosmosis, or from purely mechanical forces, as in the electroviscous effects. External fields are small—a few volts per centimeter—when the solution is a good conductor. However, when poor conductors are involved, the external field can be several thousand volts per centimeter. Internal fields due to charge located on interfaces can be very large. Ions are transported by diffusion and electromigration.

Electric fields influence fluid motion in several ways. Multiphase systems can be manipulated by electric forces acting directly on fluid interfaces. Separation processes such as electrophoresis, isotachophoresis, and isoelectric focusing depend on the action of an externally applied field. Mechanical effects due to intrinsic surface charge are always present in colloidal systems. Studies have been carried out for more than a century, and certain aspects of the subject are well understood. However, even though the subject is mature, some fundamental issues remain outstanding.

The reasons microgravity experiments may be useful stem from two factors. First, Joule heating due to the passage of current produces density gradients. These lead to buoyancy-driven convection. Simple scaling arguments show how such motions can be overwhelming. The temperature rise itself scales as the second power of the characteristic length, while buoyancy-driven motion scales as the first power of the temperature rise and the second power of the length. Accordingly, the strength of convective motions is directly proportional to the gravitational acceleration and to the fourth power of the characteristic length. Using high-viscosity fluids to reduce motion usually compromises attempts to study interesting situations. Thus, until recently, convection could be suppressed only by decreasing the length scale, and most work was done on a microscopic scale. In other situations, density contrasts are intrinsic. With two-phase systems, for example, sedimentation is always present unless isopycnic systems are used. Unfortunately, their use thwarts study of large classes of interesting materials. Here again, we are forced to work on small length scales.

From these simple arguments we see that buoyancy-driven motion makes it difficult to observe the electrokinetic phenomena in fundamental studies, and certainly influences benefits from practical applications. Microgravity experiments enable one to avoid, or at least suppress, effects of nonelectrokinetic motions.

Three sorts of problems can be cited as examples of areas where fundamental studies are in order. The first concerns the rheology of two-phase systems where imposition of an external electric field alters the constitutive behavior of the suspension. A second area involves electrically driven flows in nonhomogeneous buffers. In what is known as "isoelectric focusing," for example, the passage of a current itself alters the buffer compositions so as to cause large concentration gradients. Here interest centers on the mechanics of the flow known as electroosmosis. Microgravity experiments are almost essential if modern flow visualization schemes are to be used. The final example concerns the behavior of fluid-fluid interfaces, where experiments with electrically stressed drops and bubbles provide a means to understand the mechanics of the interface.

One of the goals of research in electrokinetics is to formulate and test theories for the behavior of electrically stressed systems. Consider, for reference, one of the classical models from continuum physics used to describe the dynamics of ordinary fluids such as air or water. The description of homogeneous fluids of this sort involves conservation laws for mass and momentum, constitutive relations connecting the internal stress to the strain rate, and specifications as to how boundaries interact with the fluid. For motions involving simple interfaces, nonlinear effects arise in two ways: through spatial accelerations—the so-called inertial terms in the equations of motion—and through the geometry of deformable boundaries. It is generally accepted that this model serves to explain a variety of situations. By solving carefully posed boundary value problems, processes ranging from the Brownian motion of small particles to the flow around certain airfoils can be described. Apparently, this model is capable of representing turbulent flows, although such situations are still mathematically intractable. Thus, although the mathematical difficulties may be formidable, the theoretical foundation is well established.

This is not the case with flows involving ionic solutions where an adequate model for the electromechanical behavior of interfaces is lacking. With fluid interfaces, charge may be transported across the interface by diffusion and migration of ions in an electric field. At the same time, the interface may dilate and contract due to motion of the bulk fluid. Any study of the interface is complicated, perforce, by the behavior of the adjacent fluids. Little is known about the nonequilibrium properties of charged interfaces. It is clear that the correct constitutive model will be nonlinear, inasmuch as the Maxwell stress tensor for the electrical force is nonlinear; but there is considerable uncertainty about the details. Experimental results are needed to guide the theoreticians and test alternative theories.

On the other hand, much has been learned about the equilibrium behavior of interfaces during the past two decades through direct measurement of the attractive (dispersion) forces and the repulsive (electrostatic) forces that make up the Derjaguin-Landau-Verwey-Overbeek theory. Theory and experiment are in close agreement. The next stage will include examination of the behavior of nonequilibrium systems, stressed by mechanical and electrical forces. Here we are on much less well-prepared ground, since theories of nonequilibrium interfaces are in the embryonic stage. It is possible that the microgravity environment can provide the venue for experiments that will enable us to understand the electrokinetic properties of interfaces to a degree comparable to that with which we comprehend their electrostatic behavior.

COMBUSTIBLE MEDIA

The central scientific issues in combustion are: flame initiation, the characteristic temperature-composition structures of flames, flame morphology, propagation speeds, and flame existence limits. The coupling of complex chemical kinetics with fluid mechanics is a general characteristic of flames. At normal gravity, buoyancy forces are observed to introduce flame asymmetries and more convoluted structures than is thought to be characteristic of the underlying transport phenomena. Take, for example, the combustion characteristics of uniform, quiescent clouds of fuel drops or fuel particulates. Gravitationally induced sedimentation processes make it difficult or impossible to experimentally establish such two-phase systems at 1 g. For both single- and two-phase flame structures, characteristic buoyancy-induced flow rates of flame gases are comparable to experimentally observed flame propagation rates in the neighborhood of flame extinction limits. This results in strikingly different limiting fuel concentrations for upward and downward flame propagation in normal gravity. The essential reasons for flame extinction limits are not adequately understood, either for premixed, gaseous flames or for premixed two-phase flame structures.

Another example is the experimental determination of flame propagation rates and flame extinction conditions for a premixed, quiescent cloud of fuel particulates in a gaseous oxidizer. In a terrestrial experiment, uniform clouds of particulates are created through use of suitable mixing processes. Quiescence is achieved only after the decay of mixing-induced turbulence and secondary flow. Downward flame propagation cannot occur where particle settling speeds exceed flame propagation rates. At 1 g, settling speeds of large particles are comparable with flame propagation rates, especially near flammability limits (the lower limit of fuel concentration for quasi-steady flame propagation). For example, unit density 50-μm particles settle at about 7 cm/s. Theories that suggest lower flammability limit flame speeds cannot be tested for the larger particle sizes. A microgravity environment would permit the establishment of a quiescent uniform cloud of premixed particulates for study.

Another issue in combustion concerns the spherically symmetric combustion of a fuel droplet in an oxidizing atmosphere. The theory aims at predicting ignition, gas-phase unsteadiness, liquid-phase unsteadiness, quasi-steady burning rates, disruptive burning, and extinction processes for the spherically symmetric case. Drop towers have been employed in preliminary studies aimed at establishing and burning nearly spherically symmetric drops. However, experimental times needed to observe the entire quasi-steady burning history of a drop vary as the square of the initial droplet diameter. Space-based microgravity environments are needed to permit the long experimental times required for the study of large drop sizes. Other combustion phenomena that are substantially distorted by 1-g conditions include autoignition of large premixed gaseous systems, two-phase combustion phenomena involving large liquid-gas or solid-gas interfaces, radiative ignition of solids and liquids, smoldering and its transitions to flaming or extinction, the structures of laminar gas jet flames, and the properties of counterflow-diffusion flames.

It is noteworthy that computational methods and combustion theory permit the analysis of combustion processes in microgravity much more readily than is possible when buoyancy is present. However, the experimental observations needed for comparison with buoyancy-free combustion theory are generally not available. These are of considerable theoretical importance since theoretical analyses for interpretation of 1-g observation generally truncate both the chemical kinetics and the dimensionality of the combustion phenomena studied. Thus, greater theoretical and computational efforts are needed to support and exploit the results of these experimental studies.

Soot formation is another area where microgravity experiments could be useful. In the combustion of hydrocarbons and metal oxides, particulate formation during the combustion of metallic particulates involves poorly understood high-temperature condensation processes. The experimental characterization of condensation kinetics generally requires the analysis of the condensation-sustaining flame structures. Flame theories have been constructed for gravity-free environments, but the ingredients require careful testing even without the complication of gravity. Inasmuch as buoyancy effects at 1 g have been shown to substantially distort counterflow diffusion flame structures, experiments in microgravity are needed to support these studies.

5
Scaling and Acceptable Acceleration Level

Throughout this report, the quantitative discussion of the effect of gravity on the phenomena involved has been in terms of a steady value of the acceleration imposed on the system. This is in part because any such estimates appropriately begin with a discussion of the effect of a steady microgravity background. But it is also due to the fact that our current detailed understanding of the scaling laws and dynamic response pertains primarily, but not exclusively, to the case of a constant, spatially homogeneous acceleration. The task group recognizes, however, that the microgravity environment is characterized by both a steady background and transient excursions in both the magnitude and orientation of the acceleration vector, and that these excursions are not necessarily small and are potentially rich in spectral content.

A concern that must be addressed when discussing the potential of microgravity research is that of acceleration levels that are necessary or desirable to meet a given scientific objective. Although much attention has been given to this question, there is no single numerical value that pertains to the wide spectrum of microgravity experiments. Indeed, the task group suggests that even within the context of a single experiment, there again is no *single* value of a minimum acceptable acceleration level. Such a value always represents a trade-off between various competing parameters that conspire with gravity to affect the phenomena under study. There will be inevitable constraints in our ability to vary these competing parameters at will. This point of view can be illustrated by amplifying some of the scaling arguments given previously.

Consider, for example, particulate systems in which fluid velocities or velocity gradients are generated by the action of buoyancy, as in the gravity settling of clouds of combustible particles, suspensions, and colloidal dispersions. The importance of such fluid motion is measured by the magnitude of the dimensionless Grashof number,

$$\text{Gr} = \Delta \rho g l^3 / \rho \nu^2$$

Thus, for sedimentation to be negligible, we must have Gr << 1. In this simple example we see that lowering g from its normal value g_0 has the *identical* dynamical effect as lowering the density mismatch, $\Delta \rho$ between the phases, reducing the characteristic length scale, l, or raising the fluid kinematic viscosity, ν. Arriving at acceptable acceleration levels, then, represents a compromise between the desired spatial extent of an experiment (which may in turn be set by the spatial resolution of instrumentation), the degree to which the chemical system of interest may be made isopycnic, and the ability to alter fluid properties. For example, if the spatial extent is to be 3 cm, working with a fluid of the viscosity of water, 10^{-2} cm^2/s, and nearly neutrally buoyant systems, $\Delta \rho / \rho = 10^{-2}$, we have

$$g << \nu / l^3 (\Delta \rho / \rho) \simeq 10^{-3} \text{cm}^2/\text{s} \simeq 10^{-6} g_0$$

Increases in length scale or density mismatch can be compensated for by either decreases in the gravitational level or increases in fluid viscosity, or both.

Such scaling arguments can and must be made in each case where a microgravity experiment is contemplated. Futhermore, they are seldom as simple as the preceding example and often involve subtleties. Consider as a second example the growth of aggregates. The geometrical properties of such aggregates may exhibit a fractal scaling behavior between a lower and upper length scale, and it is reasonable to require data over at least three decades of scale in order to make quantitative measurements of the fractal dimension. Consider an aggregate composed of particles of size 10 μm, and assume a lower cutoff of the length scale of 10 particle diameters. (Such estimates of the lower cutoff may be obtained from computer simulation, for example.) Then aggregates of macroscopic dimension

$$l \simeq (10\mu m)(10)(10^3) \simeq 10^5 \mu m \simeq 10 cm$$

might be required for accurate determination of fractal dimension. Requiring the sedimentation velocity to be small will lead to scaling arguments very similar to those given above. However, another consideration, namely the structural integrity of the aggregate, must be taken into account. Sedimentation, however slow, will exert a hydrodynamic force on the particles, which, for a loose floc, will scale in the following way:

$$F \simeq a\mu V$$

where a is a typical particle size, μ is the fluid viscosity, and V is the sedimentation velocity. Furthermore,

$$V \simeq (\Delta\rho/\rho)gl^2/\nu$$

so that

$$F \simeq a\Delta\rho g l^2$$

Knowledge of the strength of the attractive force holding the aggregate together may be used with the above estimates to set acceptable gravitational levels in order that the aggregate not be torn apart by the hydrodynamic forces. Again it should be emphasized that such acceptable gravitational levels are for a particular system, and may be altered by altering some other property of the system.

Acceptable levels of "g-jitter" or transients in microgravity may also be arrived at by use of scaling arguments. However, in order to do so we require information on the dynamical response, and in particular on the resonant frequencies of the system under consideration. This information is often lacking, and must be determined through ground-based dynamic testing, theoretical calculations, or both.

In summary, the task group finds that, without considering unusual material properties or unreasonable tolerance on density matches, levels of 10^{-5} to 1^{-6} g would allow experiments of reasonable scale to be relatively free of buoyancy-driven flows. However, the task group emphasizes that each experiment implies a separate set of requirements, and that many of them may well involve gravitational levels of substantially lower magnitude.

6
Discussion and Recommendations

In the introduction to the microgravity portion (Part B) of this report, the task group sketched a possible long-term research program in physics and chemistry in a microgravity environment and tried to state its rationale. In the succeeding four chapters this program was illustrated with examples of experiments, some of which are now in various stages of development and others of which are envisioned for later during the period covered by this report. Much of the ongoing research in microgravity physics and chemistry is being funded by NASA's Physics and Chemistry Experiments in Space (PACE) program. PACE currently supports about 15 experimental projects throughout the country. To date, none of these experiments have been flown. However, one of the experiments is about to fly, and flight hardware for others is beginning to be built. As the projects move from earth-bound preliminary experiments to full-flight experiments, the cost of each experiment increases by about one order of magnitude. Therefore, the natural maturation of the program now occurring places severe demands on program funds. Because of this, little effort is now being made to solicit new proposals. Until the funding of PACE and other basic-science microgravity programs is substantially increased, few if any new projects can be started. Even with funding available to bring these experiments to readiness for flight, the task group expects flight opportunities to be very restricted for the remainder of this decade.

Although flight facilities for microgravity experiments are now limited, the task group expects this picture to be much brighter in the era of the Space Station. Space Station planning has been, and continues to be, very responsive to the needs of experiments in microgravity. For example, the evolution of the "power tower" concept of the Space Station to the "power arrow," and on to the "twin keel" configuration has been driven largely by the needs of microgravity experiments. The Microgravity and Materials Processing Module is another example of the importance given to microgravity in Space Station planning.

The task group foresees an important symbiosis between microgravity physics and chemistry and materials processing in space. These two programs will not only share facilities, they will share technology, much of which will be developed during the next decade. For example, the task group expects the lines of research described in the preceding chapters to require more precise measurement and lower g levels from now until well into the next century. This development will place technical requirements on space systems that are difficult to predict in detail. However, it is very important for NASA to attempt these predictions now. It is equally important to plan detailed strategies for meeting these requirements and to do the necessary testing to predict the feasibility and merit of each strategy. This planning should be pursued vigorously *now*, because configurations of the large space systems that are to be in operation in the 1995 to 2015 time frame are now being determined. Decisions made now can preclude technical strategies that microgravity research in the twenty-first century will demand. Therefore, it is important to anticipate these strategies as accurately as possible.

By way of example, consider the acceleration environments that may be present in the spacecraft on which future microgravity experiments will be carried out. By acceleration environment is meant both the steady state acceleration level, g, its orientation relative to the experimental apparatus, and the time-dependent acceleration ("g-noise" or "g-jitter"). At present, the task group is reasonably confident that the noise or jitter requirements can be met at all frequencies above 1 Hz by combinations of careful noise control and properly designed isolation tables. The steady state acceleration requirements are not as easy to meet. Microgravity experiments that are now being developed for flight require literal "microgravity." They assume gravity levels down to the 10^{-5} to 10^{-6} g range. Evolution of some of these experiments into the twenty-first century will call for "nano" gravity, i.e., 10^{-9} g and below. Technology development to meet these and other requirements of the microgravity and nanogravity experiments should be vigorously pursued.

This task group believes that the microgravity environment offers opportunities for important scientific advances in the testing of fundamental theories and the discovery of new phenomena and new states of matter. In particular:

1. Our ability to test fundamental concepts such as renormalization group theory can be greatly enhanced in the absence of gravity-induced nonuniformities.

2. Our understanding of nonlinear, nonequilibrium phenomena—for example, fluid flow, condensation, solidification, combustion, and a wide range of related dynamical processes—can be significantly advanced by experiments under conditions where the underlying behavior is not masked by the effects of gravity.

3. New static or dynamic states of matter that do not exist in normal gravity may be created and investigated.

In order to take advantage of these opportunities, we must recognize certain unique features of the microgravity program: its relevance to an extremely broad range of scientific activities and its resulting need for versatility and responsiveness in its operation. Specifically, the task group recommends that:

1. NASA should increase its support for ground-based fundamental research in areas for which microgravity experiments may be relevant. These resources are to be focused on isolating those aspects of complex problems that can be uniquely investigated in space.

2. NASA should significantly expand its support of one or more centers for microgravity research. These centers should become national facilities devoted both to the basic science and to the development of the specialized skills and technologies that are needed for the effective implementation of microgravity experiments.

3. The microgravity program should include strategies to attract outstanding scientists into the program with the expectation of research successes in a reasonable time frame.

4. In scheduling flights, the basic microgravity experiments discussed above should be given the highest possible priority. In this way, the process of interplay between scientific discoveries and new commercial applications can proceed in a timely fashion.